Signals and Communication Technology

For further volumes:
www.springer.com/series/4748

Joud S. Khoury • Chaouki T. Abdallah

Internet Naming and Discovery

Architecture and Economics

Springer

Joud S. Khoury
Internetwork Research Department
Raytheon BBN Technologies
Boston, MA, USA

Chaouki T. Abdallah
Electrical and Computer Engineering Department
University of New Mexico
Albuquerque, NM, USA

ISSN 1860-4862 Signals and Communication Technology
ISBN 978-1-4471-4551-6 ISBN 978-1-4471-4552-3 (eBook)
DOI 10.1007/978-1-4471-4552-3
Springer London Heidelberg New York Dordrecht

Library of Congress Control Number: 2012950031

Printed on acid-free paper

Springer is part of Springer Science+Business Media (www.springer.com)

To my family, and to the memory of my father
—Joud Khoury

To my three C's
—Chaouki T. Abdallah

Preface

In October of 2006, as part of an NSF grant to design the future Internet, the authors got involved in designing a naming and addressing architecture for the future Internet. The Internet's naming and addressing architecture has widely accepted limitations. Its dated and host-centric Domain Name System (DNS) fails to keep up with the current and emerging usage models, and the overloading of the IP address hinders both mobility and security. As we started sifting through the large body of architectural schemes and proposals, two concerns quickly became evident. First, the independent nature of the many contributions to the field has led to the overloading of architectural terms, and to the emergence of a large body of network architecture proposals with no clear understanding of their cross similarities, and their unique properties. Second, there seems to be a growing consensus in the community on the need for designing a "smarter" network. While such evolution is bringing new potentials and service models, the community generally lacks consistent frameworks for thinking about such models and their design implications.

Purpose and Scope This book is a timely attempt at developing a unified model for studying internetwork naming and discovery. Naming is an integral building block within data networks and systems and is becoming even more so as complex data-centric usage models emerge. Several books have been devoted to the broad and general topics of computer networks [104, 154], and distributed systems [53]. We specifically focus on the naming, addressing, and discovery services. We develop the architectural and economic aspects of designing naming and discovery schemes within the broader context of internetwork architecture.

Organization The balance of this book is divided into three parts. Part I aims at framing the naming and discovery services within the broader context of the internetwork architecture, and serves as a reference for the rest of the book. Chapter 1 refreshes the reader on the main events, technologies, and design principles that led to the Internet we know today, and discusses why the very original design that enabled decades of unprecedented evolution and innovation on the Internet is falling short of satisfying the user requirements after three decades of its inception. Global clean-slate initiatives, such as NSF's FIND [10] and FIA [9], are under way to design the Internet of the future. Chapter 2 surveys some of these prominent proposals highlighting their naming aspects.

Part II presents the formal models for naming and discovery. In terms of naming, Chap. 3 disambiguates the plethora of terms used in the literature—name, address, identifier, locator, binding, routing, discovery, mapping, and resolution. We provide a formalism based on simple relational algebra, we examine the main properties of names and bindings, and we present two illustrative case studies. Chapter 4 presents a simple taxonomy that serves three main purposes. The taxonomy is a tutorial for introduction to the field, it helps in organizing and thinking about the solution space, and more importantly in identifying new avenues for contributing to the topic. In terms of discovery, Chap. 5 presents a general model for exploring the tradeoffs involved in designing Internet name-based discovery systems. Designing a discovery system is commonly challenged with providing three desirable properties: scalability, efficiency. and adaptability. The model highlights these tradeoffs across a wide range of networks. Chapter 6 is an illustrative design exercise. We present an architecture for differentiated discovery services.

Part III develops some of the economic and incentive problems common to naming and discovery services in large scale networks. Chapter 7 presents a broad treatment of the main economic issues that arise in the context of name-based discovery. Providing a discovery service while accounting for the cost and making sure that the incentives of the players are aligned is the general theme of the chapter. We motivate the subject, present a taxonomy of discovery schemes and proposals based on their business model, and we pose several questions that are becoming increasingly important as we proceed to design the internetwork of the future. This sets the stage for Chap. 8 which presents an incentive model for route distribution in the context of path vector routing protocols (mainly BGP). We model BGP route distribution and computation using a game in which a BGP speaker advertises its prefix to its direct neighbors promising them a reward for further distributing the route deeper into the network. The neighbors do the same thing with their neighbors, and so on. The result of this cascaded route distribution is an advertised prefix and hence reachability of the BGP speaker. We first study the convergence of BGP protocol dynamics to a unique outcome tree in the defined game. We then study the existence of equilibria in the full information game considering competition dynamics.

To keep the content simple and intuitive to the reader, we develop the concepts progressively and we provide examples whenever possible. The book has a logical layout to it and we recommend reading the parts sequentially as each part sets the stage for the next. For more advanced networking readers, however, we kept each chapter self-contained to facilitate skipping to particular chapters.

Audience This book is directed at students and professionals seeking a training in advanced networking topics. We believe that it may be specifically appealing for graduate students and researchers specializing in the areas of Internet architecture, naming and discovery, routing, and economics.

Acknowledgements The authors acknowledge the colleagues and the reviewers for their valuable comments, and they gratefully acknowledge the National Science Foundation (NSF) for supporting the early work investigating the field. The authors would like to emphasize that they have done their best to correct any errors that were brought to their attention. Any remaining errors are the responsibility of the authors only. Should you locate any, please email the authors at joudkhoury@gmail.com.

Boston

Joud Khoury
Chaouki T. Abdallah

Contents

Part I
Internetworks

Chapter 1
The Internet

The only constant is change
Heraclitus

In less than three decades, the Internet was transformed from a research network available to the academic community into an international communication infrastructure. How did it all start and where does it stand today? This chapter refreshes the reader on the main events, technologies, and design principles that led to the Internet we know today, and discusses why the very original design that enabled decades of unprecedented innovation on the Internet is falling short of satisfying the user requirements after three decades of its inception. Fortunately, the history of the Internet and its original design principles have been thoroughly documented in several books and articles, notably [17, 49, 80, 141], from which the material in this chapter is directly derived.

1.1 History

It is Friday October 4, 1957, when Dwight Eisenhower, then president of the United Sates, along with his soon to be appointed Secretary of defense, Neil McElroy, learns of the Soviets successful launch of the *Sputnik* satellite. The news leads to a national confidence crisis in the scientific and defense leadership of the United States. The president's response is quick. In early 1958, the Advanced Research Projects Agency (ARPA) is born with the mission of advancing military research and technology, and controlling defense research funding. Soon after, in 1961, NASA is established into law and ARPA's mission and budget are scaled back. The agency's new mission under the leadership of president John F. Kennedy is to support high-risk fundamental research. In October 1962, Joseph Licklider, a psychologist whose passion about computing quickly established him as a reputable computer scientist, establishes and heads the Information Processing Techniques Office (IPTO) at ARPA. Computers then are very expensive, bulky, scarce, slow and utilized mainly for batch processing. Licklider, however, thinking ahead of his

J.S. Khoury, C.T. Abdallah, *Internet Naming and Discovery*,
Signals and Communication Technology, DOI 10.1007/978-1-4471-4552-3_1,

time, was amazed by the potential of computers to augment human capability and simultaneously frustrated by their lack of standardization, integration, and interoperability. Collaborating with the top computer scientists in academia, a group which he refers to as the Intergalactic Network, Licklider is set to explore the possibility of networking computers, in addition to time sharing, and human computer interaction. As put in [80], "and therein lay the seed of Licklider's grandest vision yet", the ability to network disparate heterogeneous computers each acting individually. Succeeding Licklider at ARPA, Bob Taylor carries the networking idea forward and quickly convinces his leadership to allocate the money to build the first such network. After all, the defense department has every incentive to enable the sharing of computing resources across the country's research labs to minimize cost and to avoid vendor lock-in through interoperation. Bob Taylor has just the right guy in mind to run this program, Larry Roberts, a scientist at MIT Lincoln Labs whom Taylor had previously funded an experiment of his to connect two remote computers. In 1966, Roberts quickly starts establishing himself as the main architect to design, synthesize and fund the most influential ideas posited by the community, the outcome of which is the "ARPA net". At the time, two scientists have been independently working on the problem, Paul Baran and Donald Davies, of which Roberts was completely unaware. Baran's expertise in communication systems, his belief in the need for a communication infrastructure that could survive a nuclear attack (given the arms race and the anxiety with the Russians), along with the promise of digital computing, lead him to devise the two most innovative ideas of a digital computer network: survivability, and efficiency through resource sharing. Inspired by the brain's neural network's ability to survive neuron failures by circumventing them on different neural paths, the "distributed" network Baran envisions similarly avoids single points of failure. Nodes connect with redundant links to several of their neighbors so that if some node(s) fail, data may still move around failures on different paths. In addition, to avoid the inefficiencies of circuit-switching, mainly the setup delays and the end-to-end dedicated circuit resources, Baran envisions data being split into small-sized "blocks". Each block contains all the information it needs to reach its intended destination. The receiving node reassembles the data blocks to reconstruct the original message. Baran's claims are thought radical and blatantly implausible at the time and are directly opposed by phone service monopolies, primarily AT&T. Fortunately, Donald Davies, a physicist at the British National Physical Lab (NPL), arrives at a very similar network design in parallel, completely independently from Baran. Davies coined the term data "packets". Roberts then collaborates with Len Kleinrock at UCLA, Weis Clark at Washington University, Douglas Engelbart at SRI, Frank Heart at Bolt Beranek and Newman (BBN), Baran and Davies and many others to devise a formal request for proposals to build the first high speed digital network to connect up to 19 sites across the continental United States. One of the main requirements of the network is the need to decouple the routing and message processing functionality from the host computers into some other set of computers inside the network. These would come to be called the "Interface Message Processors" (IMPs)—an early version of the router. Given the scarcity of host resources and development knowledge at the sites, the host operators are opposed to allocating their host computing resources to do message processing activities (not to mention the management overhead that would partake). The IMPs are responsible for doing exactly that, reliably. Roberts presents his "ARPA net" ideas for the first time at an ACM conference in Gatlinburg Tennessee in 1967. The proposal is sent out for bids in July 1968. To everyone's surprise, BBN wins the contract to build the IMP network, expected to be up and running with four sites online within a year. Reliability is a critical requirement from the beginning. Nodes and links can fail but the network must not fail.

There are three main components that need to be built in order to deliver such a network. First, building the IMPs themselves, software and hardware. This is BBN's job. Second, the data links to connect the IMPs and the hosts are to be delivered by AT&T. Third, the host sites are expected to write site specific host-to-IMP interfaces. After all, each site has its own host computers, which are not interoperable with other hosts. Before the sites could write the host-to-IMP interfaces, BBN is expected to deliver the IMP interface specifications. Among the BBN engineers, Robert Kahn takes up the task of writing the host-to-IMP specifications. Kahn, along with the BBN engineers, decide early on that the network be completely oblivious of the hosts and the protocols they run i.e. the network exposes a standard interface to the hosts. The network's sole purpose would be to packetize messages, and route them reliably to their intended destinations. Hosts would be expected to agree on end-to-end message semantics. At the time, Vint Cerf and David Crocker who worked for Kleinrock at UCLA take on the task of designing and implementing both host-to-IMP as well as host-to-host interfaces collaborating with researchers at the other sites, a group that came to be known as the Network Working Group (NWG). The "layered" architecture is conceived as a means to logically separate the lower level transmission details from higher level application semantics. BBN builds and delivers the first IMP to the UCLA site on time around labor day and follows that with the second IMP shipped to SRI on October 1, 1969. Days after the two heterogeneous computers are networked, they successfully start communicating through a remote terminal application. In the coming years, all 19 sites are connected by the BBN built IMPs and AT&T's high speed data lines. Application protocols start to follow. The File Transfer Protocol (FTP), RFC 354, is published in mid 1972. Roberts with the help of the community and Robert Kahn, who is still at BBN, publicly demonstrate the ARPAnet by the end of 1972. At this time, a confident, proud, and passionate international networking community is emerging. In the 1970s, electronic mail (or email) becomes the killer application on the ARPAnet, the application with the most traffic; however, the network is still underutilized. The evolution of email results in intense debates that highlight the need for protocol and "header" standardization. Originally email is carried by the FTP protocol. It is not until 1982, after the invention of TCP/IP, discussed shortly, that the Simple Mail Transfer Protocol (SMTP) is rolled out as the standard email protocol.

Kahn moves to work for Roberts at DARPA and several types of network start emerging including a radio connected network and a satellite connected network. Additionally, Davies lab in London makes strides in creating a local packet-switching network, and so does Louis Pouzin in France. With all these disparate networks emerging, Kahn and Cerf, with Roberts' support, realize the need to interconnect them to broaden the possibilities beyond the ARPAnet experiment. The two men have embarked on an effort to lay the ground for what we call today the "Internet". The problem of connecting disparate packet-switched networks is not trivial. After all, each network has its own packet sizes, reliability mechanisms, and protocols. Cerf and Kahn publish their seminal inter-networking paper in 1974 [46]. The paper introduces several revolutionary designs to inter-networking. First, instead of the IMP-provided reliability that BBN built into the ARPAnet, the authors propose "end-to-end" reliability borrowing from Pouzin: the network is simply a bit plumbing medium that provides a "best-effort" packet delivery service; end hosts, using the Transport Control Protocol (TCP), are expected to perform the necessary acknowledgements and retransmissions to guarantee reliability. This enables growth and innovation at the "edges" with minimal changes to the network itself. Second, "gateways" are proposed to reside in-between heterogeneous networks with the purpose of abstracting their differences. TCP, however, is now responsible for both routing, deemed a network function, and reliability, a host function. To speed up the growth of the network, Postel posits in 1978 the need to logically separate these two functions: the Internet Protocol (IP) is responsible for packet routing while TCP is responsible for reliability. IP carries all the information that gateways need to route packets to their intended destinations. TCP resides on top of IP and handles reliability, mainly packet ordering, acknowledgements, and retransmissions. With the invention of TCP/IP, Ethernet, and the UNIX operating system, the network grows exponentially. Ethernet, invented by Bob Metcalfe at Xerox PARC labs, enables the proliferation of local area networks. Now instead of growing by adding more user accounts to a networked mainframe host, more and more hosts are added to the network. As a result, a new problem emerges with naming the different hosts on the network. Initially, each networked host maintained a map of the names and binary IP addresses of all other hosts on the network. Managing and distributing this mapping table to the hosts was centrally handled by the Network Information Center (NIC). With the rapid network growth, managing this table quickly becomes inefficient and error prone. The technical team realizes that enabling the anticipated growth on the Internet requires a more decentralized and efficient system to manage names and to map between names and IP addresses. This gives rise to the hierarchical naming architecture of the Domain Name System (DNS) designed mainly Mockapetris at USC's ISI in the late 80s. DNS enables decentralized management of the namespace. It assigns to each "domain", a collection of host computers managed by a single authority, its own subspace within a global hierarchical namespace and delegates the management of that subspace to a name server residing in the domain. In 1989, ARPAnet is decommissioned and replaced with the higher-speed National Science Foundation (NSF)-funded backbone network, NSFnet, composed of self-subsidized and commercial regional networks. The Internet enters the commercialization phase.

1.2 Architecture and Design Principles

Architecture is a very broad and overloaded term that is applied to a wide range of systems and domains from business systems (e.g. enterprises), to computing systems (e.g. information and data, software and hardware systems, technology), to economic systems (e.g. markets) and so on. Architects often struggle with evaluating and communicating the value of "good" architecture simply because the term architecture itself is not easy to define, let alone to measure. Our main goal in this section is to reflect on the original design principles of the Internet which led to its current architecture. After understanding how the Internet architecture came to be, we try to reflect on its value and limitations. To do so, however, we start by defining what we believe makes a "good" architecture and why.

The broadest definition of architecture that we find useful throughout this book is *a framework for minimizing the cost of change*. As Heraclitus puts it "the only constant is change". If we assume that a system is static, then one may argue that architectural debates become less relevant. It is only when we need to account for change that the value of architecture suddenly becomes clear. Implicitly, to minimize the cost of change, a system needs to be agile i.e. to quickly introduce and respond to change—generally the longer it takes, the more expensive is the solution. The majority of architectural patterns such as layering and decoupling, reuse, service orientation, and so on naturally satisfy this definition.

The basic design principles of the DARPA Internet are clearly outlined by David Clark [49]. Clark examines the connection between the intended goals of the DARPA Internet and the design decisions that govern its current operation. The primary goal of the Internet was to allow multiplexed utilization of its resources. Packet switching is a direct response to the inefficient resource utilization and significant delays inherent to circuit-switched telephone networks at the time. As discussed in the previous section, packet switching additionally was intended to provide survivability in the face of node and link failures, in the process of designing a network that could survive a nuclear threat. The survivability requirement later led to maintaining flow state at end hosts while keeping the core stateless with respect to ongoing flows.

Building on packet-switching technology, the requirement to connect several disparate and heterogeneous regional and remote networks resulted in making a minimal set of assumptions regarding the underlying function provided by the network which is "the network can deliver a packet". In addition, the same requirement led to the adoption of the current Internet structure which consists mainly of domains that are independent administrative authorities, and gateways (or routers) that are responsible for abstracting the differences between the domains being connected. It is more efficient and effective for a stateless network to recover from a failure. Reliability is provided by adding end-to-end connection state at the communicating hosts. Maintaining a minimalistic set of functionality at the core and implementing more specific application level functions at the endpoints is at the heart of what is known as the Internet's "end-to-end" design principles [138]. A stateless network core means the gateways are oblivious of TCP reliability. This, in addition to

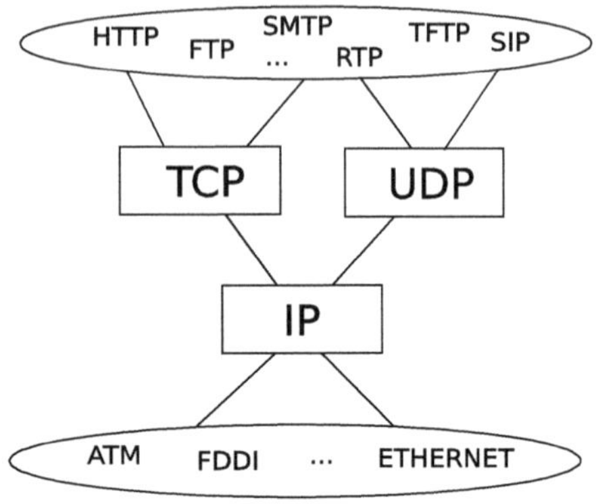

Fig. 1.1 The Internet hourglass

the requirement for alternative transport services in terms of latency and reliability, resulted in the logical split of TCP/IP. The Internet Protocol (IP), responsible for packet routing, carries only the necessary information for gateways to correctly route packets. TCP, which resides on top of IP, is an end-to-end host protocol which handles reliability.

The Internet architecture is sometimes referred to as the TCP/IP architecture or even the "hourglass". Figure 1.1 illustrates the hourglass shape with a narrow waist, represented by the Internet Protocol (IP), and a wide top and bottom. IP is the common language that all the networks need to speak in order to successfully deliver packets between them. Aside from IP at its core, the architecture assumes very little about the rest of the protocol stack above and below IP. Additionally, while the architecture itself does not dictate how to organize protocols, layering has evolved as the standard model for doing so. Layering allows for clean functional decoupling and separation of duties, and hence lower cost of change and faster evolution of the network.

This design philosophy has been central to the evolution of the Internet. Its simple and open architecture has enabled tremendous growth at the edges whether in number of hosts and users or in applications. The simple "end-to-end" principles and the low cost of entry have been major enablers of this growth. Starting with 100 hosts on the ARPAnet/Internet in 1977, that number grows by a factor of ten between each of 1984, 1987, and 1989 to reach around a million hosts in 1992 just a year after the World Wide Web goes public. There are more than a billion hosts connected to the Internet today. In the early to mid 1990s, the release of the Web's code by CERN (mainly the HTTP protocol and the HTML language) along with the invention of the browser and the passing of the Boucher act (which legalized the commercialization of the Internet) lead to a tremendous increase in the utility of the network to everyone, especially with the proliferation of the personal computer. In addition to scientists and academicians, everyone including businesses realize the great value of the Internet. Aside from adding capacity to the Internet core, all this change at the edges has been enabled at a relatively low cost and required minimal changes to the core. This has been the power of the Internet architecture.

The simple and open architecture has been a great success so far, but is it due for an overhaul four decades after its inception? There is growing consensus in the community on the limitations of the current architecture in supporting the requirements of today's users. The lack of mobility support and security are some of these limitations which we discuss next.

1.3 Limitations and the Quest for a Future Internet

Despite its tremendous success, there is a growing consensus in the research community that the Internet has architectural limitations, and that those need to be addressed in an effort to design a new generation of the Internet. Clean-slate efforts in the United States [7, 9, 10], Europe [6], and Japan [1] are under way to design the future Internet.

Among the main technical limitations of the Internet are the lack of mobility support, and the lack of security and trust (see [10, 13, 32, 37, 61]). Endpoint,[1] mobility refers to a dynamic change in the endpoint's attachment point possibly while a communication session is in progress. The Internet, and particularly TCP/IP, identifies endpoints with a tuple {*IP address*, *port number*}. The IP address then serves as the identifier of both the attachment point (a location in the topology) and the endpoint [47]. So far, the IP address has performed well as a location identifier since it inherently embeds topological information and thus fosters routing scalability under aggressive aggregation (the scalability of hierarchical routing is broken, however, due to continuous de-aggreation of the address space [115], and recent discoveries on the inefficiency of hierarchical routing over the Internet's topology [102]). When mobility is introduced however, IP loses any meaning of identity reference and degenerates into a pure routing identifier, alternatively *locator*. Coupling the endpoint identifier to the routing identifier hinders mobility and poorly identifies the actual endpoint, which exists independent of its network location or state. On the other hand, the lack of security has been attributed to limitations in both the network and the endpoint. Authentication for example is one of the main concerns of an architecture and it is hard to implement partly due to lack of identity support. More clearly, in the prevailing end-to-end model of the Internet [138], endpoints of a communication channel have no way of authenticating each other. Rather, the channel itself is generally secured using encryption for example. This leads to spoofing, spam, and many other forms of security breaches. Authentication, accountability, and trust are attractive design goals that could directly benefit from a means to identify network objects across all layers of the protocol stack. Finally, other important concerns within the current architecture include the lack of the means to detect, report, and correct errors (or in general to manage the network), the difficulty of incorporating emerging technologies and devices (sensors, vehicles, RFIDs, etc.), and the economic barriers that prevent coordination among the different stakeholders

[1] Check [138] for more on endpoints in the prevalent end-to-end design of the Internet.

(see Chap. 8 for an example). In addition to its limitations, the Internet is evolving in ways that were not anticipated during its design. As pointed out by Clark [37], the erosion of trust and the emergence of new stakeholders in the Internet are challenging the simple "end-to-end" principle which is the primary contributor to the Internet's success.

The problem of designing a future Internet is too broad. As Clark puts it [13], "The reason I stress [clean-slate thinking] is that the Internet is so big, and so successful, that it seems like a fool's errand to send someone off to invent a different one. Whether the end result is a whole new architecture or just an effective set of changes to the existing one may not matter in the end". The broadness of the scope follows because both terms "Internet" and "architecture" are too broad. The Internet is a network of networks. It is different things to different parties whether those are operators, businesses, enterprises, government, or users. These parties have different and potentially conflicting goals [37]. The tussle between accountability and privacy/anonymity, and the net neutrality debate [163] are examples of such conflicting goals. However, any architectural design must be based on a well-defined set of goals. Besides the fact that it is hard to converge on these goals, the design space is too wide to explore and too many parameters are involved. While specific dimensions of the design space have been thoroughly examined and understood, such as the TCP/IP protocol [143, 148], we still do not understand the implications of mixing different design parameters—as given by the interactions between different protocols, and systems at different layers of the stack. In fact, after surveying the literature, it became obvious that the majority of the recent architectural work explores a small set of design parameters (in the sense that it is either aimed at exploring novel usage models that adhere to a class of applications, or at directly addressing a specific limitations of the current Internet). All this is compounded by the fact that both "Internet" and "architecture" are hard to model, to measure, and to simulate, not to mention an "architecture for the future Internet" [52, 62, 68].

So how should we go about designing a future Internet? There is a debate in the community on whether clean-slate design is the right approach or whether an incremental evolutionary effort is needed instead [133]. In essence, this is a question of prioritization since clean-slate and evolutionary are different starting points in the design space. Evolutionary research is backward compatible and incrementally deployable. The proponents of evolutionary research argue using simple economics that a new technology won't be adopted unless it is able to minimize the cost of transitioning from the incumbent. Hence, instead of focusing on devising clean-slate architectures, a smarter approach would be to measure what we have, understand its flaws and economics, and devise evolutionary paths to deal with them. On the other hand, the proponents of clean-slate design argue that the Internet itself started as an experiment in the lab and that the deployability of a technology is not a good measure of the value of successful research. The value of clean-slate design lies in contributing to the science and to our understanding of networks. Many of today's shortcomings may not be solvable without architectural change: routing scalability and mobility, trust and secure identity, ability to accommodate change and programmability are some examples. In summary, the debate seems to be more of a theoretical vs. practical research debate and it seems we need to distinguish between the science itself (network science) and the application (the Internet) and be able to evolve both. Even if clean-slate design stays simply as an intellectual exercise, it adds value to the science and advances the research mission. As Henry Ford once put it, "if I had asked my customers what they wanted, I'd have built a faster horse". Customers generally lead you down an incremental path to innovation if you ask them what they want. Revolutionary ideas, however, create new economies to the innovator and to the participants. Take for example Google's auto correction feature. They watched how users search the web and how they continue searching when the result is not what they are looking for and they created a disruptive technology based on machine learning to solve that problem. Users did not ask for it, but the change created a whole new economy. Disruption is naturally followed by incremental optimization, hence both are vital and the latter is a necessary foundation.

Having discussed the limitations of the Internet architecture and the quest for a future, we proceed to briefly review some of the architectural thrusts that have emerged after about a decade of clean-slate research. The first major architectural shift lies in treating data (or information objects) as first-class network citizens instead of hosts and endpoints. This data-centric view of networking is fundamentally different from the current minimalistic and communication-centric architecture. It requires novel schemes for naming, discovering, and routing towards data instead of towards endpoints (see [16, 90, 99]). One may view this trend as taking some basic data-centric functionality that is inefficiently implemented using end-to-end semantics and making it part of the network fabric i.e. making the network explicitly aware of data objects instead of just bits. Content distribution for example would directly benefit from this new model alleviating the inefficiencies in today's infrastructure [151, 160]. The second architectural trend aims at natively supporting mobility and disconnected semantics. This is a fundamental shift from the static and connected nodes assumption of the current architecture. Supporting mobility at scale requires novel ways to naming and addressing, and presents non-trivial scalability and efficiency challenges. Delay tolerant networking (DTN) and the inter-planetary Internet are some of the efforts in this direction [5, 23] along with others [15]. Several other important trends are developing. For example, the US cyber-warfare strategy is promoting the design of secure and defensible networks with varying anonymity, privacy, and attribution properties.

1.4 Summary

This chapter reviewed the history of the Internet, its main design principles, and its shortcomings in satisfying today's usage models. As we mentioned, the literature is replete with proposals that attempt to address the identified shortcomings. Some of those proposals are incremental while others are clean-slate. The next chapter aims at exploring some of this literature with the goal of exposing the reader to the problem and solution spaces when it comes to Internetwork architecture. Given that one of the widely accepted shortcomings of the Internet is the overloading of the IP address, the literature review will additionally help us explore naming and addressing proposals for the Internet and set the stage for the discussion in the rest of the book.

Chapter 2
A Survey of Novel Internetwork (and Naming) Architectures

When all think alike, then no one is thinking
Walter Lippman

Internet architectures may be broadly categorized into either incremental or radical proposals. Incremental architectures, such as [119, 130, 158], generally aim at addressing particular limitations of the current Internet architecture through patching, while radical architectures, such as those supported by the FIND [10] and FIA [9] initiatives, tend to adopt a clean-slate approach to designing a "better" Internet, without being necessarily restricted by the current Internet model.

We classify network architectural proposals based on the service model they intend to support. This highlights a growing interest in data-centric network services. This chapter serves two main purposes: (1) to survey the literature and highlight commonalities and differences across the spectrum of solutions, and (2) to present a reference for the rest of the book, and introductory tutorial. We start in Sect. 2.1 by classifying architectures based on the service model they are intended to support. In Sect. 2.2, we proceed to overview several clean-slate architectural instances that we shall refer to throughout the rest of the book.

2.1 Classifying Network Architectures

Before discussing our classification approach, we recall some general definitions. A *computer network* is an inter-connection of computers over which information[1] flows. The *network architecture* is the conceptual design and the fundamental operation structure of a computer network. Based on these definitions, one may clearly recognize the obvious defining structures of a computer network: computers and inter-connections, communication, and information structures.

[1] Information, content, and data are used interchangeably within this chapter, unless otherwise specified, to represent data abstractions recognized by the network.

J.S. Khoury, C.T. Abdallah, *Internet Naming and Discovery*,
Signals and Communication Technology, DOI 10.1007/978-1-4471-4552-3_2,

2.1.1 Classification Approach

How do we approach the classification problem given the complexity of the design space? In other words, what should the defining element(s) of our classification model be? We start by recognizing that every design is intended to support a set of goals, which generally encapsulate the pressing needs/requirements of users.[2] Generally speaking, the design process then involves converging on a set of defining structures, and proceeding to optimize those. The outcome is an architectural design that is comprised of the following abstraction levels: (1) the *outer-architecture* represents what the network user can see. This is analogous to the network service interface or *Instruction Set Architecture (ISA)* which defines the addressing modes, the data object semantics, and the available operations; and (2) the *inner-architecture* represents the internal operation structure of the network including the low level substrate structure and the functional aspects to support the outer-architecture.

We believe that both abstraction levels provide useful and complementary insights regarding the architectural landscape. Hence, to help answer our question of what the defining element(s) of the taxonomy should be, we found it useful to classify some of the existing literature based on their supported service model (or the types of service the network provides to its users). This view has helped us in understanding the underlying goals behind an architectural design. The high-level classification, which we refer to as the service-model perspective, is briefly discussed next.

2.1.2 Service-Model Perspective

Classifying architectures from this perspective is motivated by several factors. First, the service-model approach implicitly accounts for the needs of the users relative to a network, which is the ultimate goal of any network design. For example, the Internet's simple "best-effort delivery" service model came about to satisfy a set of goals, as explained in [49], primarily allowing multiplexed utilization of resources (which led to packet switching, domain, gateways), survivability (which led to end-to-end state), etc. Second, most network architectures tend to be naturally categorized and described relative to their service models. For example, we find in the literature the "data-oriented" network architecture [99], the "delay-tolerant" architecture [5], the "differentiated services (diffServ)" architecture [35], and so on. Finally, such a classification could enable future reasoning about—and evaluation of—the degree to which a particular architecture satisfies the service requirements of the users. One such evaluation methodology based on utility was proposed in [145].[3]

[2] Within the discussion, a *user* is the general term used to abstract any entity that utilizes the network services.

[3] In [145], Shenker defines utility as the degree to which a network service model matches the needs of the network users, i.e., how good an architecture is, is measured by the happiness of its users.

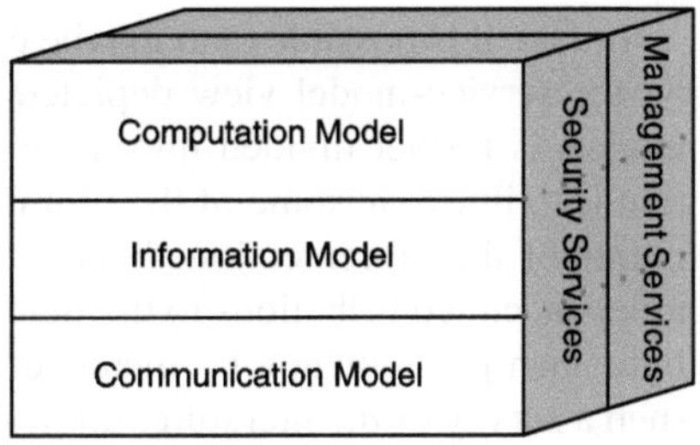

Fig. 2.1 Generalized service-model view

The generalized service-model perspective is depicted in Fig. 2.1. The communication, information, and computation[4] models represent the building blocks that collectively define, together with the security and management services, the general service model of any network architecture. By building blocks we mean that every architecture must provide these three models, whether explicitly or implicitly.[5] On the other hand, security and management services are not building blocks (since one can easily come up with architectures that do not provide any security or management services), and they operate across the communication, information, and computation models.

- *Communication model*: This service model represents the communication and control services offered by the network. For instance, delivery services whether "best-effort", QoS-aware [35], aware of disruption [5], and/or geographic location [78] all belong to this model. Communication paradigms whether connection-oriented (e.g., ATM) or connectionless (e.g., SMDS, X.25) are classified under this model as well.
- *Information model*: This model deals with the information services that the network provides to its users. The networking community currently recognizes the need for network built-in information services (naming, searching, archiving, distribution, security, and analysis services) to support a multitude of applications and their requirements.
- *Computation or Programmability model*: This model represents the level of programmability support within the network. The programmability services might potentially span all the other service models, allowing for example the programmability of the communication model and/or the information model etc. Programmable networks [44, 155], for example, provide an explicit computation model.

The security and management services provided by the network are generally, but not necessarily, offered in-band with the rest of the service models. For example, secure communication services include secure end-to-end tunneling and transport (IPSec or SSL), secure identity (HIP [118]) etc.

[4] We abuse terminology referring to the terms computation and programmability interchangeably hereafter.

[5] For example, the Internet provides an information model implicitly (the datagram and information transparency) but not explicitly.

In fact, it is possible (and maybe convenient) to fully classify the literature based on the service-model view depicted in Fig. 2.1 if each of the constituent service models is further divided into its defining elements. The major focus of this section is to illustrate some of the prominent architectural work that represents critical points of the aforementioned service spectrum. Additionally, we believe that the independent contributions to the field are converging, and this section aims to highlight such phenomenon by means of a survey. Table 2.1 presents, in matrix form, such a survey of the literature, limited to general internetwork architectures. Hence, we do not consider overlays, scoped architectural work (such as naming, or routing architectures) and we do not provide an exhaustive list of internetwork technologies which is not the goal of this section.[6] The work is divided into two parts. The first part overviews some of the independent contributions to the field, while the second part is solely concerned with the FIND [10] and FIA [9] work illustrating the community's view of what the future Internet should look like. More details about each of the architectures of Table 2.1 are provided in Sect. 2.2. Note that Table 2.1 marks the service models only as those are made *explicit* in the architectural description, and consequently it does not contradict our previous claim that communication, information, and computation models being building blocks. Several insights may be gathered by observing the matrix. First, there seems to be a growing consensus about the need for expanding the network's service model beyond the communication space, especially as researchers start thinking of designing a future Internet. Additionally, and most importantly, one can clearly notice the emphasis on information services, which is intuitive given the prevailing information-centric usage models with the current Internet.

The next section further elaborates on the service models of each of the architectures of Table 2.1.

2.2 Technical Reference

This section is intended as a technical reference for the rest of the book. We elaborate on the architectures of Table 2.1 as we refer to them frequently throughout the book.

2.2.1 Communication-Oriented

2.2.1.1 DARPA Internet

The design principles of the DARPA Internet [46] are clearly outlined in Clark's seminal paper [49] which we review here (see Chap. 1 for more history). The paper

[6]The majority of internetwork technologies (ATM, X.25, XNS, DECnet etc.) would be classified in our matrix as communication-oriented. For a comprehensive list of the internetwork technologies, we refer the reader to [48].

Table 2.1 Matrix view classification of internetwork architectures based on their explicit service-model classes, ordered by publication date

		Year	Comm	Info	Comp	Description
Independent proposals	ARPAnet [46, 49]	1970s	x			providing best-effort delivery of datagrams among globally identified attachment points
	Active Nets [155]	late 1990s	x		x	provide a framework for dynamic creation and deployment of network services at runtime
	TRIAD [77]	2000–1	x	x		exposing a "content layer" that provides transparent access and distribution of named content
	I3 [121, 149]	2002	x			providing a general-purpose rendezvous overlay that decouples host identity from location using flat, persistent names, and quasi pub-sub communication semantics
	Plutarch [54]	2003	x			provide a communication model that inherently allows interoperation of semantically disparate domains without mandating uniformity across them nor a global namespace
	FARA [50]	2003	x			provide an abstract network model that builds on the Internet's "best effort" service model adding clean separation of endpoint names from network addresses
	NIRA [165]	2003	x			providing an Internet-style communication service that inherently supports user-selected domain-level routes
	TurfNet [140]	2004	x			similar to Plutarch service model, but with global naming
	ROFL [42]	2006	x			providing an Internet-style communication service that utilizes flat labels for network identification and routing
	DONA [99]	2007	x	x		providing data access (locating and retrieving data) and data distribution independent of location using an anycast primitive and flat naming

Table 2.1 (Continued)

		Year	Comm	Info	Comp	Description
Clean-slate: FIND [10] & FIA [9]	UIA [11, 63]	2006	x			establishing communications among user devices using persistent, user-relative names and namespaces that are constructed bottom-up with no reliance on global namespaces
	Postcards [166]	2006	x	x		providing reliable delivery (push/pull) of content (large data units or files) to mobile/stationary endpoints using in-network storage/caching
	USwarm [161]	2006	x	x		providing multipoint-to-point bulk data transfer/distribution among hosts (endpoints+intermediaries) with in-network storage/caching
	ITDS [164]	2006	x	x	x	providing information transfer in response to user (endpoint) specified service expressions through in-network processing/data handling
	WiKI [39]	2006	x	x	x	providing a network query interface to users for expressing intent and implementing operations through a declarative framework for managing in-network information and state (router and host state, and data streams)
	TNA [91]	2006	x	x		provides a transient network substrate that enables identification and communication among entities based on global, and persistent (location-independent) identifiers
	PostModern [34]	2006	x		x	providing a tussle-resistant communication service, delivery of functional datagrams, that equips providers with usage control over their networks through policy enforcement, and users with policy-aware control over their traffic forwarding
	Geometric [78]	2006	x			providing a distributed location service and a spatial routing primitive for location-centric communication
	NDN [16]	2010	x	x		providing content-centric publish-subscribe primitives that allows the network to natively match interests (subscriptions) to named content (publications) by name

highlights the connection between the intended goals of the DARPA Internet and design decisions that govern its current operation. The primary goal of the Internet was to allow multiplexed utilization of its resources, which influenced by the networks (ARPAnet) and technologies (packet switching) at that time, led to the adoption of the current Internet structure: domains, packet switching, and gateways connecting networks. The survivability requirement resulted in maintaining flow state at end nodes while keeping the core stateless with respect to ongoing flows. The requirement for alternative transport services in terms of latency and reliability has led to the TCP/IP split and the introduction of UDP. The need to support various networks resulted in making a minimum set of assumptions regarding the underlying function provided by the network which is "the network can deliver a packet or datagram".

The Internet service model can be simply stated as providing *best-effort delivery of datagrams among globally identified attachment points.*

Communication Design Space

- *Topology*: The topological boundaries within the Internet are referred to as Autonomous Systems (AS) or domains. The domain is an authoritative boundary that maintains local communication policies. Domains are composed hierarchically through customer-provider or peering relationships with a distinguished core set of domains referred to as Tier1.
- *Addressing*: The Internet employs a global hierarchical addressing architecture, IPv4. The address space is composed hierarchically to match the topology, rendering the network address a topological forwarding directive. An IP address is formed of two parts: the network address and the host address.[7] The address space is finite.
- Naming: DNS naming is an out-of-band service that is not part of the core network services, i.e., the network does not recognize a domain name. Hence, the Internet model does not constrain the naming architecture design space. Whether a centralized, global, hierarchical naming architecture (DNS style), or a distributed, flat naming architecture (e.g., OpenDHT) ends up being deployed is irrelevant to the Internet architecture as long as their exists a mechanism to translate a name to an IP address.
- Routing and Forwarding: The Border Gateway Protocol (BGP) [132] is the de-facto standard for inter-domain routing on the Internet. BGP is a policy-based path vector protocol which empowers domains with control over route selection and propagation. Forwarding is thus based on the policies of the domains and the protocol allows for arbitrary preference functions over routes. While the core forwarding function on the Internet is a best-effort delivery service, guaranteed delivery is provided as an end-to-end service using TCP.

---

[7] Classless Inter-Domain Routing (CIDR) separates the two parts at arbitrary boundaries using the forward slash notation as in 192.168.1.1/16. This means the first 16 bits of the 32 bit address, 192.168, designates the network address.

Information Design Space The only information abstraction recognized by the Internet architecture is the datagram. More complex information models must be composed out of the simple datagram. The type of content being delivered over the Internet's best-effort service whether a static file or an time-sensitive voice stream is irrelevant to the architecture.

2.2.1.2 Plutarch

Briefly, Plutarch [54] proposes a framework for next generation networks that embraces heterogeneity within and handles it through contexts and interstitial functions. Contexts are like ASs that implement their internal addressing, naming, routing, and transport mechanisms. Interstitial functions map between the set of functionalities of different contexts. Plutarch is a response to the shortcomings of the current Internet protocol model that unifies all underlying network types through a one-size-fits-all networking and addressing mechanism. This (i.e., IP) has resulted in a semantic bottleneck that is leading to ossification of architecture where it is increasingly difficult to introduce anything but incremental changes. Network Address Translators (RFC 1631), and Resilient Overlay Networks [24] are examples of unclean solutions adopted as a result.

The service model of Plutarch, what users expect from the network, is not addressed in the paper. However, the aim of the work is to *provide a communication model that inherently allows interoperation of heterogeneous networks and mechanisms without mandating uniformity across networks (or a global namespace).*

Communication Design Space

- Topology: heterogeneous networks (contexts); boundary: a context is a region of the network that is homogeneous in some way, "a set of bindings with reference to which names may be resolved"; Composition: composition of contexts is not hierarchical, it is either adjacent (border) or containment (nesting); there is no notion of a global or unique "root context"; different namespaces can exist per context; no overlap of contexts.
- Naming and Addressing: local naming and addressing within contexts; inter-context name translation at gateways through "interstitial functions"; Lookup: route-query to chained-context-descriptor mapping (out-of-band); distributed route-query search across contexts (e.g., flooding).
- Routing: route over the discovered chained-contexts.
- Compatibility with the Internet: Plutarch subsumes the Internet as a context.

2.2.1.3 TurfNet

The proposal TurfNet [140] is a network architecture for future dynamically composable networks. The architecture is based on the concept of composing autonomous heterogeneous networks, referred to as turfs, dynamically without sharing a global addressing mechanism or network protocol. Composition of networks takes two forms: (1) merging or horizontal composition, and (2) hierarchical independent or vertical composition. Within each AS (turf), independent addressing, routing and resolution mechanisms (control plane functionality) are available and are local to that turf. Across the turfnets, no common network protocol and no shared address space is required. However, a global name space is required to allow communication between the turfnets.

The service model of TurfNet is similar to Plutarch except for the designated global namespace.

Communication Design Space Separate naming/addressing:

- Topology: hierarchical ASs or turfs; composed horizontally (merging) or vertically (customer-provider/peering).
- Addressing: local address space per turf.
- Naming: Global namespace; Lookup: name to address mapping (in-band, i.e., lookup creates forwarding state), recursive lookup creates address and protocol translation state (soft state) within boundary routers up to root ASs.
- Forwarding/Routing: local forwarding within turf using local addresses; intra-turf: hosts external to the turf are mapped to turf address space; mapping soft state maintained by boundary gateways; end-to-end path composed of up-path to root (created during lookup) and down-path to destination turf (created through registration); routing = address + protocol translation at boundary gateways.

2.2.1.4 Internet Indirection Infrastructure (I3)

Internet Indirection Infrastructure (I3) [149] is a general-purpose indirection overlay network over IP that uses rendezvous servers where clients can register triggers using persistent names. It decouples the sending and the receiving actions where clients send traffic to the overlay and the latter takes care of forwarding the traffic to other interested clients that registered triggers in the system. A trigger is a simple binding of a persistent name to an IP address. In addition to mobility, several communication primitives are enabled as a result including unicast, multicast, and anycast by registering different types of trigger. Traffic in I3 goes through the overlay indirection servers thus degrading the routing efficiency in contrast to direct routing over IP, and potentially causing triangle routing and doubling the amount of traffic. The simple service model of I3 makes it vulnerable to a multitude of security attacks. The proposal Hi3 [121] is an attempt at making use of an overlay (I3) in conjunction with a secure direct end to end approach mainly the Host Identity Protocol (HIP) [118]. This would allow for efficient end-to-end traffic with HIP (mobility, multi-homing, DOS resistance) that leverages an independent secure, integrated rendezvous infrastructure (I3) as an overlay to route the HIP control traffic (stability).

The I3 service model may be stated as *providing a general-purpose rendezvous overlay that decouples host identity from location using flat, persistent names, and quasi pub-sub communication semantics.*

2.2.1.5 FARA

FARA [50] is an abstract network model in which the current Internet architecture is generalized and remodeled to enable clean separation of endpoint names from network addresses. The FARA abstraction recognizes communication among pairs of entities via logical links referred to as associations on top of a communication substrate. An entity in FARA is the endpoint of communication and smallest unit that can be mobile, such as a process, a thread, or a cluster of devices. An association is a logical communication link between a pair of entities representing persistent communication state. Entities maintain local association state and may have multiple concurrent associations. A packet belongs to one association and carries an association ID (AID) that enables the receiving entity to correctly demultiplex the packet to its association. As to the communication substrate, it represents underlying infrastructure that is able to deliver packets on behalf of associations. Addressing, routing, forwarding are mechanisms employed by the substrate and are left unspecified by FARA. However, FARA assumes connectionless point-to-point communication between entities. An entity supplies the substrate with a packet and header that contains a destination Forwarding Directive (FD). The latter contains enough information that the substrate can use to deliver the packet all the way to the destination entity that contains the association. Separating entities from the substrate can lead to seamless mobility of entities and independent evolution of applications and substrate mechanisms.

In M-FARA instantiation, and unlike the traditional Internet, no global address space is assumed, i.e., no IP addresses. However, there is a set of local address spaces referred to as "addressing realms". Topological information is represented in terms of these private realms that a packet will traverse in transit towards its destination. The FD here is a set of sub-FDs specifying the realms on the path. In the case that the source entity and the destination entity belong to the same realm, the destination FD has NO topological information. When the source and destination belong to different realms, the destination FD must specify the topology which can be very complex if the private realm addressing is flat. M-FARA's addressing introduces a 2 level hierarchy with a designated globally known "core". Thus the destination FD in this case will consist of (FDup, FDdown) pair of FDs specifying how to reach the core from the source and then how to reach the destination from the core. This design supports mobility across realm boundaries.

2.2.1.6 NIRA

NIRA [165] is a novel inter-domain routing architecture that equips users with the ability to choose domain-level routes. The user control over the sequence of providers (ISPs) that packets take introduces competition among the backbone ISPs, thus driving innovation and lowering service cost. NIRA's support for user choice introduces several problems including practical provider compensation, route discovery and representation, and security, which are addressed in the paper.

The service model of NIRA [165] may be stated as *providing an Internet-style communication service that inherently supports user-selected domain-level routes.*

Communication Design Space NIRA reuses many of the Internet's design decisions:

- Topology: strictly hierarchical ASs (domains) with distinguished core (customer-provider and peering relations); concept of domain hierarchy from edge user to core referred to as up-graph.
- Addressing: globally unique, hierarchical address space (IPv6 reused); address encapsulates topological location and provider hierarchy; (scheme: provider rooted hierarchical addressing).
- Routing: valley-free; unicast: 2-segment route (uphill + downhill), one from source to core (uphill) and one from core to destination (downhill); downhill discovered through name service; proactive detection of route failures; BGP-like inter-domain routing within the core, and path vector inter-domain routing over provider hierarchy.
- Naming: naming system required to map endpoint names to downhill route segment; name system design not mandated (may be DHT or DNS style for example).

2.2.1.7 ROFL

Routing On Flat Labels (ROFL) [42] addresses the challenge of how to scalably route on flat labels with no location identifiers. In contrast to the previous proposals, ROFL eliminates location identifiers altogether, and solely depends on persistent identifiers to route inter and intra-domain traffic. The advantages of such approach in addition to mobility and multi-homing include fate sharing (no control path since no need to contact resolution infrastructure), simple address allocation, better capture of identities at network layer, and independence from any external resolution systems.

The service model of ROFL [42] may be stated as *providing an Internet-style communication service that utilizes flat labels for network identification and routing.*

Communication Design Space ROFL eliminates the address dependence on location, hence the network address itself becomes the name.

- Topology: hierarchical ASs; AS up-graph required (domain's provider hierarchy).
- Addressing/Naming: Namespace: flat circular namespace similar to Chord [150]; Name semantics: unique persistent and global identifiers, semantic free as in [29], self-certifying (HIP [118] public key hash); Naming system: global DHT formation and maintenance as nodes join and leave.
- Routing/Forwarding: compact routing; no name/address resolution since routing on flat DHT; ID translated into source route during forwarding and route follows successor pointers; DHT formed over routers and static hosts.

2.2.1.8 Postmodern Internet

The postmodern Internet architecture [34] is a reaction to the rigidity of the current Internet's network layer with respect to different and possibly conflicting policies of stakeholders. The architecture aims to provide a minimalist network layer that anticipates tussle, and accommodates for it through flexibility in introducing policies. Users need to control how their traffic is delivered, while providers try to control their network usage. Some example policies include inter-domain routing policies, packet filtering policies, policies of who can specify forwarding and who has access to what, service policies, etc.

The service model of the architecture can be stated as *providing a tussle-resistant communication service, delivery of functional datagrams, that equips providers with usage control over their networks through policy enforcement, and users with policy-aware control over their traffic forwarding*.

Communication/Computation Design Space

- Topology: hierarchical virtual realms (trust boundaries).
- Packets and forwarding: functional datagrams (smart packets containing functional blocks—how, what, where, knobs and dials); user control over forwarding paths/directives (FDs) when aligned with provider policies; (mechanisms: Resolution from "destination specs" to LinkIDs to reach destination).
- Addressing and Routing: both decoupled from forwarding and not restricted (not part of network layer), i.e., no global addressing mechanism required; (mechanism: use globally unique Link IDs instead of node IDs to determine paths; inter-realm LinkID routing information dissemination).
- Transparency: in-network packet processing and rewrites of functional blocks.

Information/Security Design Space Functional packets recognized by network; network entities (hosts and realms) have trusted identities generated by decentralized PKI infra; signed packets; accountability and path signatures.

2.2.1.9 Geometric Stack

The proposal [78] calls for making geographic/spatial location information an inherent service of the network, for the latter to better address the needs of dense wireless/mobile access networks (geographic routing/addressing/tracking/dissemination). A novel stack is proposed to provide communication through physical space rather than network space (topology) enabling a multitude of applications that utilize geo-routing, geo-casting, and localization.

The service model of [78] may be stated as *providing a distributed location service and a spacial routing primitive for location-centric communication*.

Communication Design Space Packet-switched; location information available over some coordinate systems with translation among them:

- Topology: hierarchical topology; high-speed wired backbone with wireless edge networks; nodes associated with home areas.
- Addressing: address is geographic identifier (unicast: host ID + last position; geocast: set of coordinates of a zone).
- Forwarding and Routing: linear geometric routing (along 1-D paths called trajectories); source node specified path equation and network decides on forwarding through next-hop local forwarding by intermediate nodes (not source routing but path specification).
- Naming: host identifier (e.g., MAC derivative); Lookup: distributed service to resolve identifier to location (DHT based), location-based service.

2.2.2 Information-Oriented

The architectures we discuss in this section aim to achieve a similar objective, which is efficient network information support (typically in-band), a service that the current Internet model lacks. While sharing the general goal, the design decisions (and mechanisms) employed by each proposal remain different, which is what we try to illuminate next. There is a direct coupling between the information and the communication models, and the latter is generally intended to provide the necessary requirements for efficient data access (time, space, disruption, disaster, etc.).

2.2.2.1 TRIAD

The original Internet architecture provides transport mechanisms that are transparent to the applications or services employing them [49]. As the Internet usage models become more content-oriented (web traffic, multimedia, or p2p traffic), more intelligence is overlaid on top of the traditional Internet design to provide faster and more reliable content access as is the case with Content Distribution Networks (CDN). TRIAD [77] is a novel architecture that treats content as first-class, shifting the communication paradigm from host-centric to content-centric communication. The proposal exposes the limitations of current content distribution models, whether scalability, latency, or architectural openness and consistency. TRIAD addresses the content problem by making explicit a *content layer* that can efficiently *route* towards content. The content layer spans the core of the network by extending traditional IP routers to support *name-based routing*.

TRIAD's service model can be stated as *exposing a "content layer" that provides transparent access and distribution of named content.*

Information Design Space

- Type of information: datagrams/packets, services, and content.
- Naming: semantics: URL names compatible with DNS, URL split into domain name of content server and file name; service names are persistent whereas content names (service name + file name) is not; namespace: global hierarchical namespace; naming system: distributed naming infrastructure with no single point of failure.
- Routing: name-based routing: inter-domain content routers (CRs) route based on names towards content servers (caches for closest replicas, transformers); routing is a distributed in-band search operation; single-source multicast support.
- Tussle: Content routers (CRs) are provided as ISP infrastructure (similar to BGP routers now) and are thus controlled by the domain's authority, hence ISP control over directory service. Additionally, coupling naming and routing at the domain level can potentially lead to more tussle.

Communication Design Space Host-to-content; Delivery of content depends on the communication model (TRIAD reuses HTTP/TCP/IP transport). TRIAD does not mandate the Internet model though.

- Addressing and routing: Addressing is global hierarchical (IPv4 reused). Address is only a forwarding directive used for transient routing/forwarding of information and not for lookup.
- Naming and routing: There is an explicit separation between name and address/forwarding directive. All network entities (hosts) are identified by names (DNS). Name is end-to-end identifier, information model handles routing based on names.
- Transport: in-band lookup and connection setup/transport (using DRP as an alternative to TCP), hence symmetric search and delivery paths (lookup is as available as delivery).
- Mobility: Indirection-based host mobility, Name-based routing abstracts topological location, but since endpoints (hosts) do not advertise their names (for scalability), endpoint acquires new name in visiting domain and inserts a redirection in its home domain.
- Compatibility with Internet: highly compatible with Internet infrastructure, requires extensions at directory level.

2.2.2.2 DONA

The Data Oriented Network Architecture [99] is a clean-slate redesign of the naming and resolution mechanisms on the Internet. Similar to TRIAD [77], DONA is a reaction to the evolution of the Internet usage models. Initially, the Internet was designed to enable host-to-host communication (FTP, telnet where source explicitly specifies address/locator of destination) over a transparent forwarding engine. This model has significantly changed since into a data-centric model where users access content and services independent of the location of content, services and of the users. DONA proposes replacing DNS names with flat, self-certifying names and replacing the name resolution mechanisms with a name-based anycast primitive that lives over IP. The main design requirements for DONA are persistence, availability and authenticity. The architecture itself is a synthesis of ideas from HIP [118], TRIAD [77], and SFS [114].

DONA's service model can be stated as *providing data access (locating and retrieving data) and data distribution independent of location using an anycast primitive and flat naming*. The design decisions of DONA are very similar to those of TRIAD, except for the naming architecture. We simply point the differences.

Information Design Space

- Naming: Semantics: persistent, self-certifying, flat name (HIP [118]).
- Tussle: ISP control over user's activity (content lookup and registration activity); ISP physical control over Resolution Handlers (which are similar to content routers in TRIAD).
- Security: Authentication and integrity of information (mechanism: PKey cryptography).

2.2.2.3 USwarm

Universal Swarm (USwarm) [161] is a proposal that applies swarming techniques (such as BitTorrent) to design a universal data transfer architecture that learns from p2p architectures and eliminates selfishness of peers through incentives. The architecture is a response to the data transfer shortcomings on the current Internet. The whole Internet is modeled as a single swarm that employs a distributed metadata resolution system to resolve data-to-peers(s) that can serve the data (analogous to BitTorrent Tracker). An intentional naming system is employed as well to resolve intentions to metadata. Hence 2 level resolution is proposed: intention to metadata to provider peers.

The service model of USwarm can be stated as *providing multipoint-to-point bulk data transfer/distribution among hosts (endpoints + intermediaries) with in-network storage/caching*.

Information Design Space

- Type: datagrams, data, metadata.
- Naming: Bittorrent model—semantics: self-certifying (publisher, hash) tuple, principal name globally unique and authentic, metadata uniquely specifies data and contains name plus block ids; infrastructure out-of-band intentional resolution system IRS (map intent to metadata); distributed p2p lookup/search.
- Routing/search: locality-aware distributed tracking that involves peers and intermediaries to locate content (i.e., resolve metadata to peers); more control over routing decisions for ISPs (traffic engineering) and users.

- Transport: multipoint to point transport of information; incentive-aware.
- Security: authenticity and integrity of data.

Communication Design Space Internet style point-to-point communication employed.

2.2.2.4 UIA

The User Information Architecture (UIA) [11, 63] is a clean-slate redesign of the naming and resolution mechanisms on the Internet. The current Internet reliance on global naming makes it hard for two devices to communicate even when they are in close proximity. As a result, instead of relying on global namespaces and infrastructure-based connectivity, UIA constructs user-local namespaces bottom-up and embraces ad-hoc connectivity for communication among user devices. Users assign personal names to their devices and to other devices they meet. These user-specific, persistent names constitute the user's namespace and need only be unique within that user's namespace. UIA provides mechanisms for namespace creation, name resolution, and local and remote connectivity among the users' named devices. Communication could rely on ad-hoc or infrastructure-based networks depending on the physical paths available between the named devices.

The service model can be stated as *establishing communications among user devices using persistent, user-relative names and namespaces that are constructed bottom-up with no reliance on global namespaces.*

Communication Design Space

- Naming: devices are identified with flat cryptographic identifiers (EIDs); users assign persistent, user-relative names to their devices; the user-relative names are bound to the device EIDs and form the user's namespace; each user's namespace contains user-relative names/bindings of the user's devices and of other users' devices that are merged with the user's namespace when devices are introduced; the user namespace is organized as a tree with the user as its root and other users' spaces forming subtrees.
- Routing: several options are proposed for routing on EIDs whether using ad-hoc gossip-based discovery protocols and routing for devices on a local network, or using the Internet for remote connectivity, or even using clean-slate DHT-based routing similar to other identity-based routing proposals; given that users, devices, and user-relative names are the main currency of UIA, and given that devices are linked following social introductions, UIA proposes using social structure to expedite routing on EIDs.
- Security: EIDs are cryptographic hashes of public keys; name-based authentication and access control using distributed PKI infrastructure; security is built bottom-up to follow the namespaces.

2.2.2.5 Postcards from the Edge

The proposal "Postcards from the Edge" [166] is a clean-slate cache-and-forward architecture for a future internet. The architecture is a response to the revolution in access technologies, primarily wireless and mobility, that overwhelm the Internet's basic design assumptions. Given the continuously increasing capacity and decreasing cost of in-network storage, the authors propose an architecture that provides uniform reliable transport of large files across heterogeneous access networks and in the face of intermittent connectivity.

The service model can be stated as *providing reliable delivery (push/pull) of content (large data units or files) to mobile/stationary endpoints using in-network storage/caching*.

Information Design Space

- Type of info: large data units (files).
- Naming of content: Semantics: globally unique (UFID.FQDN) names for files; name service: out-of-band service (i.e., lookup and transport are not coupled), hierarchical name resolution system (DNS style); Lookup: distributed name-to-cache(s) mappings.
- Routing/search: no routing based on names; out-of-band search/lookup to name service (i.e., lookup name then contact host similar to Internet); Rendezvous push/pull through post-office nodes (every node knows PO current PO nodes where former can pick its content).

Communication Design Space Host-to-host; builds on top of Internet best-effort IP service for addressing and routing of control traffic. IP is not essential to the design though.

- Topology: hierarchical topology formed of high-speed wired backbone connected to edge access networks.
- Naming of endpoints: location aware rendezvous service; Lookup: distributed name to post-office mapping; Security: endpoint associated with home autonomous system for authentication when mobile.
- Routing: hop-by-hop routing on location information; Supports Type-of-Service (TOS) to distinguish between transport and caching.
- Mobility: artifact of rendezvous.
- Compliant with Internet: builds on IP, hence highly compliant.

2.2.2.6 Information Transfer and Data Services

The proposal ITDS [164] is a response to the simple store-and-forward model of the current Internet. The architecture calls for in-network support of a broader range of services by dealing with information abstractions[8] rather than simple bit transfers. Hence the communication model proposed is aware of information rather than simply being a bit-plumbing medium. Additionally, in-network data processing is proposed to implement dynamic user service requirements.

The service model of [164] may be stated as *providing information transfer in response to user (endpoint) specified service expressions through in-network processing/data handling*.

Information/Data and Computation Design Space

- Type: data, information; semantics: various transfer characteristics (streaming, random access, interactive, canned, ...).
- Processing: general-purpose computation model on routers to support data services; data services already offered by network and not dynamically deployed by users (contrast to active nets for example); service specifications are mapped to computation model rather than dynamically introduced; service composition, control.
- Storage: in-network storage/caching possible on routers.

Computation Design Space User specifies intent and network maps computation to resources, hence, limited user control.

Communication design space Can operate on top of various communication models (e.g., Internet); ITDS focuses on information transfer models "on top of the network".

2.2.2.7 WiKI

The proposal Wireless Knowledge Infrastructure (WiKI) [39] addresses the limitations of the current Internet in supporting the needs of applications and services given the huge proliferation of wireless, mobile, and ubiquitous computing. WiKI takes a clean-slate approach to designing a future declarative network in which in-network state (router, network, and host state) is treated as distributed data that can be queried by users through declarative languages. Such approach separates logical representation from actual implementation making the network more flexible to change, and more informed about its operations. Hence, WiKI realizes a knowledge plane [51] for the Internet.

The service model of WiKI may be stated as *providing a network query interface to users for expressing intent and implementing operations through a declarative framework for managing in-network information and state (router and host state, and data streams)*.

[8]The proposal explicitly differentiates between information (e.g., "the requested web page did not change") and data/content (e.g., the actual web page data).

Information Design Space

- Type: internal network information (state), data streams, continuous/static queries; cross layer, cross domain views of data streams; archives.
- Processing: integration, aggregation, fusion, joins, etc.
- Security: policy enforcement; distributed monitoring; access control.

Computation Design Space Built-in distributed WiKI runtime (proxies); query processing and optimization; user control: user has control over computation (e.g., route selection/protocol, service composition) through queries (mechanism: Network Datalog language to specify computation).

Communication Design Space Wireless/mobile endpoints query WiKI proxy nodes (infrastructure or overlay) that perform query processing; can operate as overlay.

2.2.2.8 NDN

Motivated by the content-centric usage models on the Internet, content-centric networking [90] or Named Data Networking (NDN) [16] by Jacobson et al. recognizes content, instead of hosts, as a first-class network citizen. The NDN architecture introduces new content-centric primitives into the Internet's core; primitives for content discovery, retrieval, and caching allow for more efficient, secure, and location-independent information sharing. NDN re-architects the protocol stack to route interest packets towards named data objects. Interest packets are routed by name using simple name-aware forwarding and caching primitives.

The service model of NDN may be stated as *providing content-centric publish-subscribe primitives that allows the network to natively match interests (subscriptions) to named content (publications) by name*.

Information Design Space

- Objects: two types of packet, interest packets (subscriptions) and data packets (publications); data packets consume interest packet; matching on longest prefix name as in IP.
- Naming: content names are multi-part and hierarchical; names are globally routable and are aggregated by prefix in a manner similar to IP for scalability.
- Security: to provide integrity and self-authentication, each content packet is individually signed along with its name binding.

Communication Design Space

- Routing: can operate as an overlay on top of IP; only interest packets are routed towards the content provider, while content is symmetrically forwarded on the reverse path; supports multicast primitives and disconnected semantics.

2.2.3 Computation-Oriented

The most notable work on programmable networks is perhaps the Active Networks. A detailed survey on active networks is provided by Tennenhouse [155], and a more general one on programmable networks is provided by Campell [44]. The area of active networks has extensively explored the idea of programmable networks with the ability to introduce change into networks dynamically. Some pointers to prominent work in this field can be found at DARPA's site http://www.sds.lcs.mit.edu/darpa-activenet/, and at the IEEE issue [3, 12]. This section reviews an example of programmable networks instead of surveying the literature.

In [156], the authors introduce the concept of programming the network by shifting the computing paradigm on current networks from an end-to-end system to a system in which each fine-grained data element participates in the computation. Within active networks, traditional packets are replaced by "capsules" which contain programs and content simultaneously. Nodes on the network (routers, switches, servers—firewall) can dynamically execute the capsule programs safely and efficiently. The capsules are loaded into a transient environment on the node and are safely executed/interpreted eliminating the ability of the capsule programs to stray beyond the restricted environment and thus to compromise the shared resources on the network. The programming abstraction provided by active networks allows user-driven customization of the infrastructure to enable faster deployment of new services. It also enables for adaptive protocols on the network, thus tailoring the infrastructure for user/application needs. Logically, active networks shift the intelligence in the network from the node to the capsule. A capsule for example will compute its path within the network (might need to access routing tables on nodes) instead of the capsule (packet) being dumb. PLAN [85] is a functional programming language for active networks used within packets. It is resource limited and secure enabling a smart means of communicating between nodes. PLANet [84] is an active network implementation that utilizes PLAN to implement network layer on top of an IP-free link layer. PLANet uses active packets and allows active extensions, used to operate the network, to be downloaded to routers (for example to implement services like DNS, address resolution, routing). An active packet within PLANet needs to explicitly specify the destination of evaluation, avoiding the evaluation of the packet program on every node on the packet's path. This is mainly due to the reason that evaluation is computationally expensive. Programs in packets are marshalled at the source node and unmarshalled only at point of evaluation. In general a packet is only forwarded by an intermediary node (by executing an *routeFun* attribute specified by the packet) until it reaches the intended destination on which it is evaluated. Addressing within PLANet is based on 48 bit addresses (implemented with 32 bit IP appended to 16 bit port number) assigning one address per network interface on a node.

2.3 Conclusion

This chapter presented a survey of novel network architectures. We have shown that while the communication structure is necessary for defining and representing a modern network architecture, it is in general insufficient. Information and computation structures are other building blocks that need to be properly understood within modern networks. We highlighted naming as a core service within an internetwork architecture and we reviewed the basic objects and abstractions involved. As we shall see in the next chapter, most of the naming architectures are composed of simple underlying objects and primitives.

Part II
Naming and Discovery

Chapter 3
Naming

> What's in a name? That which we call a rose
> By any other name would smell as sweet
> *Shakespeare*, Romeo and Juliet (II, ii)

Names make it easier to identify and refer to people and things around us. Restaurants in a city, postal mailboxes on a street, digital files on a computer, and physical interfaces on a router are examples of things we assign contextual names to. In general, the association we make between the name and the named person or thing is a long-term one. The name "John Smith" identifies a person and the name "100 Newbury Street Apt 1F, Boston MA 02115, USA" identifies an apartment mailbox in Boston. Both names generally have a long-term association to *what* they name. A name is inherently a relative concept: different people might assign different names to the same thing. In addition to the association between the name and the named person or object, we generally make associations between objects themselves. This type of association is referred to as the *binding*. For instance, we say that "John Smith" lives at "100 Newbury Street Apt 1F, Boston MA 02115, USA". This binding between the person's name and the name of their mailbox (i.e. postal address) is a shorter term binding that changes more frequently. Now, in order to send John a letter, it is convenient to be able to specify *where* to send the letter to i.e. the destination mailbox address. This requires that the sender knows the association between the named person and the named mailbox. The sender may look up this binding in some directory service. The operation by which the letter travels to its final destination is provided by the postal service. The postal service has structured the mailbox names and the delivery centers in such a way to make it efficient to deliver letters to their destinations at increasing resolutions using a form of hierarchical naming: first deliver to the state, then to the zip code, then to the street/apartment. During this process, letters are aggregated by destination for efficiency. If instead of specifying a destination mailbox address, the sender of the letter were to just address it to "John Smith", there is a chance the letter might still be successfully delivered to John. However, the process of delivering such a letter would likely be inefficient. Assigning names to mailboxes and associating tenets to the mailboxes makes the delivery process much more efficient and much more likely to succeed.

J.S. Khoury, C.T. Abdallah, *Internet Naming and Discovery*,
Signals and Communication Technology, DOI 10.1007/978-1-4471-4552-3_3,

Computer networks are no different. We use names so we can identify, share, and locate *objects* on a network. Networks host objects, including for example nodes, endpoints, services, content, and users. A computer network is essentially intended for delivering and sharing information. The basic primitive needed for successful delivery of information is the ability to discover *route*s to objects. So in essence, the two most basic abstractions in a computer network are objects and routes. This in a sense defines the purpose of naming in the context of computer networks: to make it easier to discover routes to objects. Almost every networking application relies on naming services, the latter being an integral part of a network architecture. Revisiting the definitions of Shoch [146], Hauzeur [81], and Saltzer [137] and the elegant formulation of Karsten et al. [92], Sect. 3.1.1 starts by disambiguating the plethora of terms used in the naming literature—*name*, *address*, *identifier*, *locator*, *binding*, *routing*, *discovery*, *mapping*, and *resolution*. We present the definitions and we provide a formalism based on simple relational algebra. The model adds clarity to the discussion by formally representing the main abstractions, their relationships, and the main constraints on them. Such a formalism may be directly fed into symbolic model checkers (such as [2, 89]) to explore and check a variety of system level properties. Sections 3.2, 3.3, 3.4 then examine the properties of names, bindings, and discovery. Two illustrative case studies of name services are presented in Sect. 3.5 before concluding the chapter.

3.1 Naming and Forwarding

3.1.1 Names, Addresses, and Bindings

Semantic confusion about naming and addressing in communication networks (with the terms name, address, and route) dates almost three decades back. Shoch [146] constructed the general and simple definitions of the terms name, address, and route. Building on that, Saltzer in his RFC [137] explains the confusion by shedding a different and very helpful perspective on the constructs of naming and addressing in data communication networks. Before we revisit these definitions, we broadly introduce two types of network object that we shall refer to throughout the rest of the book: information or *Data Objects* (DOs), and communication or *Switching Objects* (SOs). A DO is the general abstraction we use to refer to any concrete network object that participates in the communication process, including for example a webpage, an electronic mailbox, a server, some protocol code in a network stack, and an interface card. SOs are a subset of the DOs that additionally have internal switching logic (and potentially switching state). SOs actively participate in message forwarding [92]. Hereafter, we use the general term *object* to refer to either type i.e. when the type distinction is not important. Objects establish relationships/associations with other objects. A relationship that is particularly important for communication is the *binding*. A binding is defined as an association between an object (whether DO or SO) and an SO. We require an SO in the binding relation since switching logic and

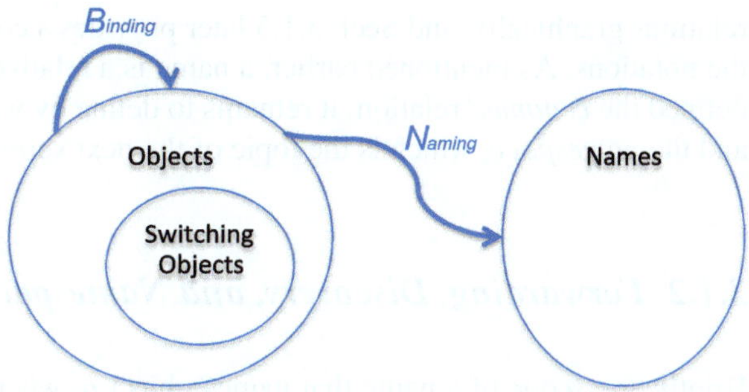

Fig. 3.1 The universe of objects $\mathcal{O}$ on a network, including switching objects $\mathcal{S}$, and the universe of names $\mathcal{N}$ and the relationships between them

state, a property of the SO only, is needed for the binding to be useful for communication as we shall see later in the discussion. A good example of an SO is a role in a role-based architecture [40], or a protocol layer in the Internet layered architecture. We now proceed to define the terms name and address.

Building on the simple definitions of Shoch [146], Saltzer [137] defines the *name* of an object as an identifier of "what you want". On the other hand, the *address* of an object o is merely a name of the object that o is bound to, hence the importance of the binding concept when distinguishing a name from an address. The important point here is that the address is just another name that we semantically differentiate from a name only after we identify the objects being named and the binding between them. Saltzer in his model argues that in order to distinguish names and addresses on the Internet, it is helpful to distinguish four different kinds of objects: a service, a node, an attachment point, and a path. These may be thought of as switching objects (SOs) using our terminology. According to his definition, the address of a node object is just the name of an attachment point object to which the node is bound, and the address of the attachment point is just the name of a path to which the attachment point is bound, and so on. The address then identifies "where the object is".

Let us add some formality to these definitions. We define the following sets: the universe of data objects $\mathcal{O}$ of which the universe of SOs is a subset $\mathcal{S} \subseteq \mathcal{O}$. We define the universe of names $\mathcal{N} = \{0, 1\}^*$ to be arbitrary binary strings. We define the following binary relationships on the sets: the first relationship is the binding denoted as $B \subseteq \mathcal{O} \times \mathcal{O}$, and pronounced *is bound to*. More clearly, we say that object o *is bound to* o', denoted as oBo' if the ordered pair $(o, o') \in B$, $o, o' \in \mathcal{O}$. As we shall define shortly, the set of bindings we are interested in requires that at least one of the ordered pairs $(o, o') \in B$ to be a switching object. The second relation we define is the naming relation denoted as $N \subseteq \mathcal{O} \times \mathcal{N}$, and pronounced *is named*. We say that object o *is named* n i.e. oNn if the ordered pair $(o, n) \in N$, $o \in \mathcal{O}, n \in \mathcal{N}$. Note that under N an object may have multiple names and a name may be associated with multiple objects (as in multicast). The inverse relation N^{-1} contains the transpose of all ordered pairs, is pronounced *names*, and has the same cardinality as N. Note that the binding relationship carries to the names i.e. when two objects are bound it is natural to refer to the binding of their names. A name alone is generally of limited value; we are interested in the name only in relation to the object(s) it names. Figure 3.1 shows the naming and binding binary relations graphically, and Sect. 3.1.3 later provides a concrete clarifying example of the notations. As mentioned earlier, a name is a relative concept and while we have defined the *is named* relation, it remains to define by whom i.e. the *scope* of a name and the *namespace*, which is the topic of the next subsection.

3.1.2 Forwarding, Discovery, and Namespaces

Briefly, the *scope* of a name that names object o, where $(n, o) \in N^{-1}$, is the set of all SOs that acknowledge this relation between n and o. As we shall define shortly, these are the SOs that when presented with a message with name n in its header, end up forwarding the message (on its way) towards o. A *namespace* is then defined relative to the *scope* and is simply the set of all names valid within that scope. Let us now formally introduce the forwarding relation along with the definitions of scope and the namespace.

We quickly introduce two relevant relational operators and some notation first:

- $\circ_k$ is the k-join of two d-ary relations, $(d > k)$: the relations are joined on k attributes.[1] For any two relations A, B, $(a, b, c) \in A$ and $(b, c, d) \in B \implies (a, d) \in A \circ_2 B$. When $k = 1$ we omit it for simplicity of notation;
- $*_k$ is the transitive closure applied to a d-ary relation R, $d > k$: basically $R^{*_k} = R \cup R \circ_k R \cup R \circ_k R \circ_k R \cup \cdots$ containing R and its transitives;
- $\tilde{A}$ denotes the transpose of relation A i.e. the ordered pairs of A in reverse order.

Now define the relation *forwards* of quadruples as $F \subseteq \mathcal{S} \times \mathcal{N} \times \mathcal{O} \times \mathcal{N}$ as a simplified abstraction of the switching logic and state of all SOs in the network. An ordered quadruple $(s, n, o, n') \in F$ where $(s, o) \in B$, $s \in \mathcal{S}$, $o \in \mathcal{O}$, and $n, n' \in \mathcal{N}$ means that when presented with some *message* with name n in its header (name of the intended target), SO s forwards the message to object o rewriting the name n in the header to n'. Note that forwarding in this model is not restricted to communication over some network link. It includes forwarding between collocated SOs on the same node as in the IP layer SO forwarding an IP packet to the MAC layer SO to which it is bound for example. By definition F represents the global forwarding function and F^{*_2} (the transitive closure) is simply the *routing* or more generally the *discovery* function[2] i.e. a quadruple $(s, n, o, n') \in (F^{*_2})$ if and only if (i) o is reachable from s, formally $(s, o) \in B^*$ (there is a binding path between the origin SO and the destination object), and (ii) each SO u along the path from s to o knows how to correctly forward the message given some name.[3] We shall distinguish routing from discovery later in Sect. 3.4. For now, we define the reachability function

[1]The last k attributes of the first relation must equal the first k attributes of the second in order for the rows to survive the join.

[2]Discovery/Routing additionally includes the distributed control mechanisms needed to construct and maintain the forwarding tables.

[3]By definition of F, an object $o \notin \mathcal{S}$ such that $(s, n, o, n') \in F$ must be terminal.

$reach(s,n)$ as the set of objects that are reachable by the forwarding function F^{*2} when a message with name n arrives at SO s. Formally,

$$\begin{aligned} reach(s,n) &= \{o \in \mathcal{O} : (s,n,o,.) \in F^{*2}\} \\ &= n \circ (s \circ (F^{*2} \circ \mathcal{N})) \\ &\subseteq \mathcal{O} \end{aligned}$$

The *scope* of a name that names object o is then given by the following function that takes as input an ordered pair $(n,o) \in N^{-1}$ and returns the set of all SOs that can reach o when presented with name n, formally

$$\begin{aligned} scope(n,o) &= \{s \in \mathcal{S} : o \in reach(s,n)\} \\ &= (((F^{*2}) \circ \mathcal{N}) \circ o) \circ n \\ &\subseteq \mathcal{S} \end{aligned}$$

and the scope of a name n is,

$$\begin{aligned} scope(n) &= \{s \in \mathcal{S} : reach(s,n) \neq \phi\} \\ &\subseteq \mathcal{S} \end{aligned}$$

A namespace relative to some scope is the set of all names defined within that scope, or formally

$$\begin{aligned} namespace(scope(n)) &= ((scope(n) \circ F) \circ \mathcal{N}) \circ \mathcal{O} \\ &\subseteq \mathcal{N} \end{aligned}$$

A *naming service* manages and implements the naming relation N. Finally, notice how we defined an ordered quadruple $(s,n,o,n') \in F$ whereby SO s maps a name n to n' during forwarding. This process of the mapping between names is referred to as *name mapping* or alternatively *name resolution*, and is generally part of the naming service. In its simplest form, the SO may keep a static local table of name mappings or it could alternatively utilize some service to do that dynamically, as we describe later Sect. 3.1.3.

Having formally defined the notions of name, address, and binding, let us put these definitions to test. When trying to classify the Internet Protocol (IP) address as a name of the attachment point or a name of the node on the TCP/IP Internet, confusion may arise. Saltzer [137] works through the following example to try to clarify the confusion: consider a node x having IP address $x.y$. The node changes its attachment point and keeps the address $x.y$ unchanged in the new attachment point by changing the forwarding tables within the network. Is the IP address then the name of the node? Saltzer argues that, while one may be tempted to think that the IP address is a name of the node since it remained unchanged across changes of the attachment point, that is not the case. He argues that by changing the forwarding tables within the network, what has really happened is that the permanent name of the new attachment point has changed instead. Hence, Saltzer claims that the IP address is the name of the attachment point object.

Let us try to answer the question using the formal definitions. First, assume the IP address of the node is public and is part of some provider aggregated prefix. In order to see which object is being named by some name n, the IP address in this case, we start by evaluating $scope(n)$ which yields the set of SOs aware of the name n. In this case, this contains the layer 3 SOs (IP layer) of all nodes on the network. This should be a good indicator of the object being named, in this case, it is the IP layer of the node, which we denote by s. Consider two scenarios: in the first scenario where the node relocates to a different provider and changes its address to n', this means that the new IP address n' now *names* the node's IP layer s i.e. a new association $(s, n') \in N$ between the name and the named object exists and $scope(n, s) = \phi$ after relocating. In the scenario where the node maintains the same IP address n after relocation by changing the forwarding tables within the network,[4] then clearly we can claim that the name of the IP layer s did not change i.e. $(s, n) \in N$. In summary, while in both scenarios the binding of the node's MAC layer SO changed due to relocation, the name only changed in the first scenario and one could claim that by changing the network forwarding tables, s could retain the name association (s, n). The key insight here is that the validity of the name n (its scope) is defined relative to the forwarding function. There is a subtle difference when it comes to changing the name of an object versus changing the binding which generally confuses the discussion. One needs to be careful to distinguish the two types of relation introduced here: the relation *named* between a name and the named object (which is generally but not necessarily a long-term one); and the binding of an object to another object.

3.1.3 Illustrative Example

To better illustrate the definitions of names and bindings, let us examine the different names and bindings involved when a browser attempts to retrieve a resource on the Internet, say a webpage with the name http://www.unm.edu:80/index.html. Figure 3.2 depicts this process graphically. We show four nodes, each distinctly labeled. The HTTP client on node 1 is requesting a webpage/resource from the HTTP server on node 4. The webpage name is referred to as the Uniform Resource Identifier (URI) and it basically names a file object o on the webserver, index.html. First we list the universe of objects. We denote the SO on node i by $type_i$ where *type*

---

[4]For example using a provider-independent address that is de-aggregated and that each node in the DFZ has a record for it in its forwarding table.

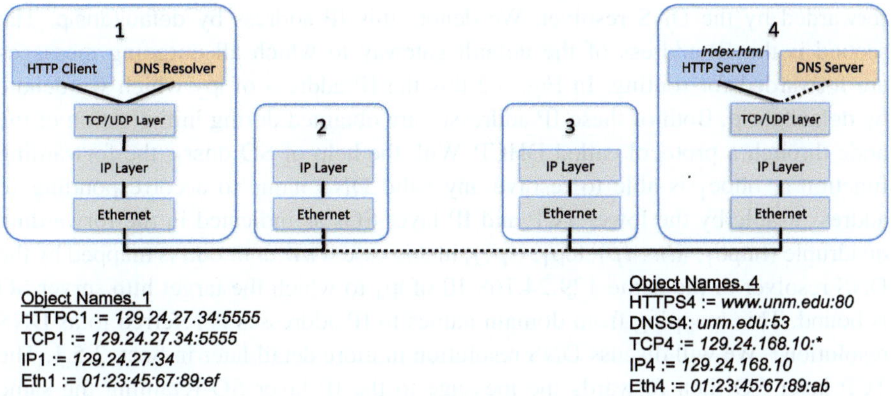

Fig. 3.2 Example of names and bindings on the TCP/IP Internet. We placed the DNS server SO (in the *unm* domain) on node 4 for simplicity of presentation

takes values in {httpc, https, dnsr, dnss, tcp, ip, eth}; for example, httpc_1 denotes the HTTP client SO on node 1. The universe of objects consists of 15 SOs in this example, the set $\{type_i\}$ along with the file object index.html. The universe of names has the same size, where each SO is named with a unique name; Fig. 3.2 shows the names of the objects on nodes 1 and 4. The *is named* relation N contains 15 ordered pairs; for example, $(\mathsf{https}_4, \mathsf{http://www.unm.edu:80}) \in N$. The *is bound to* relation B is bidirectional and is given by

$$B = A \cup \tilde{A}$$

where the binary relation A comprises the following pairs:

$$\begin{aligned} A = \big\{ & (\mathsf{httpc}_1, \mathsf{tcp}_1), (\mathsf{dnsr}_1, \mathsf{tcp}_1), (\mathsf{tcp}_1, \mathsf{ip}_1), (\mathsf{ip}_1, \mathsf{eth}_1), \\ & (\mathsf{ip}_2, \mathsf{eth}_2), \\ & (\mathsf{ip}_3, \mathsf{eth}_3), \\ & (o, \mathsf{https}_4), (\mathsf{https}_4, \mathsf{tcp}_4), (\mathsf{dnss}_4, \mathsf{tcp}_4), (\mathsf{tcp}_4, \mathsf{ip}_4), (\mathsf{ip}_4, \mathsf{eth}_4), \\ & (\mathsf{eth}_4, \mathsf{eth}_3), (\mathsf{eth}_3, \mathsf{eth}_2), (\mathsf{eth}_2, \mathsf{eth}_1) \big\} \end{aligned}$$

We list some of the forwarding quadruples for the SOs on node 1, it is straightforward to derive the rest. These include: $(\mathsf{httpc}_1, \langle dns|ip\rangle, \mathsf{tcp}_1, \langle ip\rangle)$, $(\mathsf{dnsr}_1, \langle dns\rangle, \mathsf{tcp}_1, \mathsf{defaultdnsip})$,$(\mathsf{tcp}_1, \langle ip\rangle, \mathsf{ip}_1, \langle ip\rangle)$, $(\mathsf{ip}_1, \langle ip\rangle, \mathsf{eth}_1, \langle eth\rangle)$, and so on, where the notation $\langle.\rangle$ means some value and | lists options.

Going back to our example, the HTTP client httpc_1 is trying to retrieve object with name www.unm.edu : 80/index.html. The name www.unm.edu is a domain name in the Domain Name System (DNS). During initialization, there are two important names that are discovered/configured on the SOs on node 1: the first is the IP address of a local DNS service to which DNS resolution queries are initially forwarded by the DNS resolver. We denote this IP address by defaultdnsip. The second is the IP address of the default gateway to which all outgoing messages are forwarded for routing. In Fig. 3.2 this the IP address of ip_2 which we denote by defaultgtwip. Both of these IP addresses are obtained during initialization of the node through a protocol called DHCP. With the help of SO dnsr_1, the forwarding function at httpc_1 is able to resolve any valid DNS name to a corresponding IP address usable by the lower TCP and IP layer SOs as indicated in the forwarding quadruple $(\mathsf{httpc}_1, \langle dns|ip\rangle, \mathsf{tcp}_1, \langle ip\rangle)$; in this case www.unm.edu is mapped by the DNS resolver to the name 129.24.168.10 of ip_4 to which the target http server SO is bound. This mapping from domain names to IP addresses is referred to as DNS resolution.[5] We will discuss DNS resolution in more detail later in Sect. 3.5.1. The TCP layer SO then forwards the message to the IP layer SO retaining the same destination IP address 129.24.168.10. The IP layer SO then determines that the message needs to be forwarded to the default gateway (default next hop) which has IP address defaultgtwip. With the help of the Address Resolution Protocol (ARP), the defaultgtwip IP address is mapped to an Ethernet address, Eth2 in this case, usable by the Ethernet SO eth_1. This mapping is indicated by the forwarding quadruple $(\mathsf{ip}_1, \langle ip\rangle, \mathsf{eth}_1, \langle eth\rangle)$. The recursive process of name mapping and forwarding continues until the message reaches its final destination.

Notice how $scope(129.24.168.10, \mathsf{ip}_4) = \{\mathsf{httpc}_1, \mathsf{tcp}_1, \mathsf{ip}_1, \mathsf{ip}_2, \mathsf{ip}_3, \mathsf{ip}_4, \mathsf{tcp}_4, \mathsf{https}_4\}$ contains both httpc_1 and tcp_1 in addition to ip_1. This is what we shall refer to later in Sect. 3.5.2 as the overloading of the IP address as it names two objects simultaneously: the endpoint object tcp_4 (endpoint of a TCP connection [47]) and the IP layer object ip_4 (attachment point in [137]). Also notice how the names here have different structures and how the name of the target HTTP server SO www.unm.edu : 80 is independent of its current binding. One could have named the http server with 129.24.168.10 : 80 instead and saved the DNS resolution overhead during the forwarding process. While it speeds up the forwarding (see Sect. 5.2.3 for more on Internet routing), the name 129.24.168.10 : 80 embeds location and makes it harder for the named SO to relocate (change its binding). We shall discuss such properties of names and bindings in further detail in the next two sections.

3.2 Properties of Names

We have so far introduced the definition of the name, the naming relation, and the namespace. In this section, we discuss how names are chosen. We distinguish all names of network objects along two dimensions: *syntax*, and *semantics*.

[5] The nslookup utility on windows or unix systems is a DNS resolver.

3.2.1 Syntactic

In terms of syntax or structure, names come in different forms. The most common distinctions that have been made in a network setting include: human-readable vs. binary, and multi-part vs. flat. As their name suggests, human-readable names are meant to be used by humans. These are generally employed at the interface between humans and the network [30]. Email addresses and the domain names from which they are derived are good examples of human-readable names. In general, these names are organized to reflect human organizational names and structures as with domain names (discussed in more detail later in Sect. 3.5.1). On the other hand, binary names are intended to be used by computers and need not be human-friendly. IP and MAC addresses are good examples of binary names from the Internet. IP addresses are used by Internet routers to refer to other routers while MAC addresses are used in Ethernet interface cards/drivers to refer to other cards.

The other common syntactic distinction is *multi-part* vs. *flat* (or single-part) names. Multi-part names are composed of multiple parts each of which may be human-readable or binary. *Hierarchical* names belong to this class and are attractive as a structural and decentralized way to provide global uniqueness of names. Names need to be unique only within the same "prefix" of the hierarchy but they can repeat across prefixes. Domain names are a good example of hierarchical names. IP addresses are another example. The 32-bit IP address 129.24.27.10/24 is composed of four 8-bit parts, which represents a concatenation of the network address (in this case the first 24 bits, 129.24.27, as indicated with the /24) with the host address (in this case the last eight bits, 10). On the other hand, user routes are an example of multi-part names that are not necessarily hierarchical. These may be used to identify end-to-end paths on the network. In contrast, flat names are made of a single part and are void of structure. Flat names are generally less attractive for humans especially at large scale. Examples here include flat random bit strings such as with binary cryptographic hashes. Hashes, for example, additionally guarantee uniqueness within a large universe with high probability. We describe such a scheme in Sect. 3.5.2.

3.2.2 Semantic

Regardless of their syntax, names may or may not embed certain semantics. At the highest level, we distinguish these two classes of names: *semantic-free* vs. *semantic* names. Semantic-free names are void of semantics in the sense that they are independent of the content or attributes of the objects they name. These names are attractive as *persistent* identifiers: they remain unchanged even across changes in the content or attributes of the objects they name/identify. A good example of persistent identifiers are Uniform Resource Names (URN) or handles in the Handle System [8, 153].

On the other hand, semantic names come in a variety of forms depending on the content or attribute semantics they embed. We classify those names by distinguishing the types of attribute that network objects generally have. These come in three main classes: communication, information, and computation attributes, and accordingly semantics. A good example of an object's communication attribute is its network location. The topological IP address on the Internet, or the coordinates in some metric space [126] are examples of location-based names. Location-based names are referred to as *locators*, and are attractive as forwarding directives [45]. Examples of an object's information attributes include content ownership and type as with domain names. The name www.unm.edu embeds the content type (www for web) and the domain ownership of the server (unm). Content identity and integrity is another example as with cryptographic names [99, 107, 150]. Finally, examples of computation attributes are intermediate transformation utilities for the object as with active names [159].

3.3 Properties of Bindings

Bindings exist between objects and evolve over time. The most common properties of bindings distinguished in the literature are *temporal* properties. These describe the times at which the binding gets created and destroyed.

3.3.1 Temporal

Bindings that do not change over time are referred to as *static* bindings, while those that do are termed *dynamic*. Static bindings are attractive in that they may be cached over long periods of time instead of having to be discovered every time they are needed. When two objects are statically bound, then the association between their names is also static; a web service SO s_1 with name www.unm.edu : 80 that is statically bound to an IP layer SO s_2 with name 129.24.168.10 implies that the association between the two names may be statically stored within the network without having to map/resolve from one name to the other every time during the forwarding process. This increases the efficiency of forwarding at the cost of brittleness or lack of adaptability when the bindings change. Should the web service bind to a different SO after relocating, then it would become unreachable using the static binding. Indeed, this is the case with the current Internet naming of TCP endpoints which does not survive mobility.

Dynamic bindings on the other hand are bindings that change more often over the lifetime of the objects. These are more attractive in highly dynamic environments where objects move around. Dynamic bindings generally add overhead to the forwarding process as a result of name mapping but may still benefit from caching for fixed periods of time over which the bindings might be static. Two related terms have been coined in the context of dynamic bindings: *early* vs. *late* bindings. Early vs. late has to do with the time at which the binding between the objects (and consequently their names) is fixed during the forwarding process. When it is fixed at the source, then it is referred to as early binding. In our example, when forwarding a message to SO s_1 with domain name www.unm.edu : 80, the name is resolved through DNS to IP address 129.24.168.10 at the source before the message is forwarded; hence, the binding between the target object s_1 and the object s_2 it is bound to is fixed at the source. In highly dynamic and frequently disconnected environments, this binding might change during the forwarding process invalidating the binding at the source. Late binding instead discovers the binding incrementally during the forwarding process. This has the advantage of adaptability at the cost of potentially further reduction in efficiency. The Delay-Tolerant Networking (DTN) architecture (RFC 4838) [5] is a good example where late binding is very useful.

3.4 Discovery

Before introducing the properties of discovery, we first clarify the subtle distinction between *discovery* and *routing*. Discovery is the end-to-end process that allows a message to reach its intended destination object. During the process of delivering the main message, several other messages might need to be exchanged for successful delivery. As we have seen in the example of Sect. 3.1.3, in order to deliver some main message payload $\mathcal{M}$ to www.unm.edu : 80, a separate DNS resolution message had to be sent by the DNS resolver to map name www.unm.edu to IP address 129.24.168.10, and several additional ARP messages had to be exchanged. The main message $\mathcal{M}$ along with the DNS and ARP messages are all necessary for successful delivery and they all comprise the discovery process. The DNS and ARP messages do not contain the main message payload and hence are not part of the routing. Routing is only the subprocess of the discovery that acts on the main message $\mathcal{M}$ i.e. the process by which the main message $\mathcal{M}$ is directly forwarded. In our example, routing starts after the DNS name is resolved and the IP address is known. In this sense discovery subsumes routing. We shall use the general term discovery hereafter and make the distinction only when needed.

The design of discovery schemes aims at satisfying a set of requirements and is based on a set of assumptions. Some of the common requirements we observe in the literature include *efficiency*, *scalability*, *adaptability*, *user-control*, *robustness*, *resilience*, *manageability*, *trust*, *security*, *privacy and anonymity*, *accountability*, *economic requirements*, etc. In terms of assumptions, the most common ones address the underlying graph structure, and model. For example, assumptions about the graph model include general ones such as hierarchical, scale-free, or small-world assumptions, or more specific structural assumptions of underlying metric embeddings. Other assumptions specify whether the graph is static or dynamic. In order to highlight some of the inherent tradeoffs in the design of discovery schemes, we explore several architectures and we examine the tradeoffs between scalability, efficiency, and dynamism support in detail in Chap. 5.

3.5 Naming Services Case Studies

A naming service, as defined in Sect. 3.1.2, is responsible for managing and implementing the naming relation N. In general, this involves assigning, storing, and mapping names. In this section, we discuss two architectures of naming services driven by different requirements on the names. The first architecture is based on hierarchical names while the second architecture is based on flat names. The two case studies highlight very different requirements, properties, and challenges.

3.5.1 Domain Name System

As mentioned in Chap. 1, in its early days, the Internet had only a few hundred hosts. In order to keep track of the IP addresses of these hosts, a centralized table of hostname to IP mappings was kept in a *hosts.txt* file that was centrally managed by the Network Information Center (NIC). The file was manually updated every so often and distributed to the sites. The community quickly realizes that this manual process is error prone and results in inconsistencies. This leads to the birth of the Domain Name System (DNS) [117].

The global DNS namespace is organized hierarchically starting with a set of root names (suffixes) that include: com, org, net, edu, gov, and mil, in addition to one per country such as us, uk, lb, and so on. Under each suffix, subspaces of the namespace are created and delegated to *domains*. A domain is nothing but some entity that manages a subtree of the namespace. Generally, domains coincide with human organizational structures (businesses, universities, clubs, etc.). ICANN, the authority that oversees domain names, has been actively trying to expand the global Top Level Domains (gTLDs) beyond the ones listed above to admit arbitrary gTLDs in different languages. As of the writing of this book, large scale adoption of such gTLDs is still unclear due to lack of demand, and incentives.

The hierarchical nature of the domain namespace provides a simple structural means to guarantee global uniqueness of names on the Internet. Another key aspect of DNS is the decentralized management of the namespace. A domain that owns a subtree of the namespace completely manages the subtree including the names and the database servers that serve the names within the subtree. In our example, UNM owns the unm.edu subtree i.e. all possible names suffixed with unm.edu. UNM also manages the authoritative DNS servers that serve these names.[6] This means that DNS servers (which serve/resolve the names) themselves are organized hierarchically to closely follow the namespace tree. The database managed by a DNS server is simply a file, referred to as the *zone* file, that comprises a set of *resource records* for the domain. Figure 3.3 shows the hierarchical structure of the domain name servers coinciding with the edu tree and shows an excerpt from an example zone file for domain ece.unm.edu.

---

[6]UNM could delegate the management of its namespace to some third-party instead, but it is still the authority of the subtree.

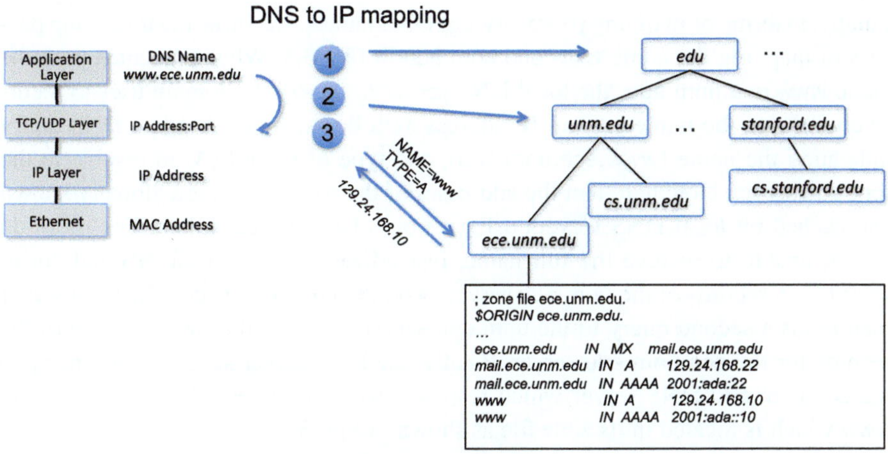

Fig. 3.3 The TCP/IP stack is shown to the left with the namespace used within each layer noted next to it. A (sample) subtree of the domain server hierarchy is shown to the right. Mapping an application Layer DNS name (www.ece.unm.edu) to an IP address involves three queries shown with the arrows in the middle: first contact the top level edu domain server in the hierarchy, then the unm.edu server, then the ece.unm.edu server. We show an excerpt of the *zone* file located on the latter which shows the IPv4 and IPv6 entries for the www name

DNS was originally intended to name Internet hosts. While it may theoretically name other objects, the most widely accepted record types refer to different types of host (servers). Each records stored on a DNS server has a type which specifies the type of object named in the record.

- Type A: is an IP version 4 address. As shown in the zone file in Fig. 3.3, the type A record corresponding to the name www is the IP address 129.24.168.10 i.e. a type A query for name www.ece.unm.edu yields this IP address.
- Type AAAA: similar to type A except this is the IP version 6 address.
- Type NS: names the authoritative servers that are responsible for resolving a prefix within the domain. For example, in Fig. 3.3, the umm.edu zone file would generally contain an NS record for ece with the name of the server that manages the sub-domain ece.unm.edu. The NS records are generally accompanied by their corresponding A (and/or AAAA) records.
- Type MX: names the mail exchange servers within the domain. The MX records are generally accompanied by their corresponding A (and/or AAAA) records. Figure 3.3 shows the MX records in the ece.umm.edu zone file. These are the servers that mail programs contact when delivering mail to john.smith@unm.edu for example.
- Type SRV: names the Session Initiation Protocol (SIP) [135] servers within the domain. The session initiation protocol is used for a wide variety of media applications on the Internet (such as voice-over-ip). The SRV records are generally accompanied by their corresponding A (and/or AAAA) records.

Finally, in terms of mapping (or resolving) DNS names, we show the three-step process of mapping the name www.ece.unm.edu in Fig. 3.3. When presented with the name www.ece.unm.edu, the local DNS server in the origin domain tries to recursively resolve the name into an IP address as follows: first it sends a DNS query indicating the name (www.ece.unm.edu), and type of record (A) to resolve to the root edu server. Note here that the addresses of the root servers are globally known and cached by local DNS servers using an out-of-band mechanism. The root edu server, unable to resolve the full name, instead responds with the NS and corresponding A records of the unm.edu servers who own that subspace. The local server then sends a second query to the unm.edu servers who do the same and return the records for the ece.unm.edu server. Finally, the local server sends a third query to the ece.unm.edu DNS server which responds with the A record corresponding to www which is located in its zone file as shown in Fig. 3.3.

3.5.1.1 Properties

We have already mentioned the two fundamental properties of DNS: *global uniqueness* through hierarchical organization, and *decentralized management* of the namespace and name service through delegation. These two properties together foster wide adoption as they eliminate the barriers to entry and minimize the need for coordination among participants. New domains join the service at minimal cost and without the need for any coordination with other participants. Several other attractive properties arise as a result.

Decentralization provides a *simple trust model*: a service that is owned and operated by a domain is authoritative for that domain. In other words, I am likely to trust the association between www.ece.unm.edu and 129.24.168.10 simply because the domain that owns the name is serving this association.[7] In addition, decentralization results in *aligned economic incentives with no externalities* whereby each domain bears the cost of their namespace, and adding a new domain to the global hierarchy does not increase the cost on other participants.

Fate sharing results from the fact that global availability of a domain service is only dependent on its ancestors in the hierarchy, and is completely independent of other domain servers that are not ancestors. Additionally, local availability is completely independent of any other service (including ancestors) which means that machines within a local domain may still resolve local domain names regardless of the state of the rest of the hierarchy. Finally, hierarchical structure with caching of names results in scalable and efficient resolution.

---

[7] As long as the DNS protocol is trusted to behave correctly.

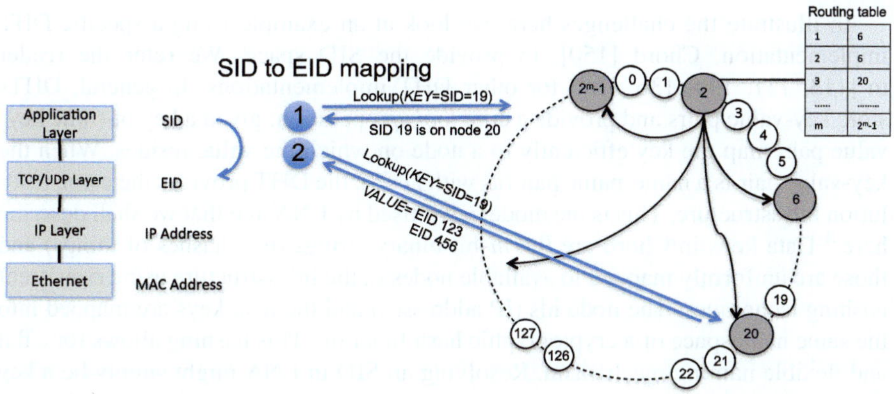

Fig. 3.4 The TCP/IP stack is shown to the left with the namespace used within each layer noted next to it. A (sample) Chord DHT [150] is shown to the right for m-bit binary strings organized on a ring. The larger shaded circles represent actual physical nodes in the DHT. The Chord routing table for node 2 is also shown (finger table); node 2 keeps pointers, black arrows, to all available nodes at positions following $2 + 2^j \bmod 2^m$ (there is a logarithmic number of those). Mapping an application Layer SID i.e. a key (in this case 19) to a value (in this case a set of EIDs) involves two queries shown with the arrows in the middle: first contact any node in the DHT with the *lookup* query. The response is the identifier and IP address of the node that stores the requested SID (the node closest to the SID on the ring, in this case node 20). Then a second query is sent to node 20 which returns the EID values associated with the SID

3.5.2 Layered Naming

The Layered Naming Architecture (LNA) proposal [29] is a response to the overloading of the IP address and the shortcomings of the DNS system in meeting the demands of an ever-expanding information network. Motivated by the increasing applicability of Distributed Hash Tables (DHT), the LNA proposal is among the first to combine flat, and semantic-free names for the TCP/IP Internet. Briefly, LNA is based on the following simple principle: use flat and persistent names for objects at different layers of the stack. If two objects can exist independently, then they must be named independently as well. If instead one object's name is derived from another's based on their binding, then unwanted complexities arise when the bindings change. As we saw earlier in our example of Sect. 3.1.3, the TCP layer object is named using the same name as the IP layer object. This association of the names causes issues when the binding changes; the TCP connection fails. As a result, the authors advertise the use of two new namespaces: the first namespace is for services or data objects (application layer objects) and is referred to as Service IDs (SIDs), while the second names TCP objects (transport layer) and is referred to as Entity IDs (EIDs). The authors suggest using flat, semantic-free names for both namespaces to provide persistent identity. Accordingly, two new name mapping/resolution services are needed, the first maps from SIDs to EIDs, and the second maps from EIDs to IP addresses as shown in Fig. 3.4. DHTs are proposed as a scalable solution for providing flat name storage and resolution.

To illustrate the challenges here, we look at an example using a specific DHT implementation, Chord [150], to provide the SID space. We refer the reader to [113, 131, 134, 136, 168] for other DHT implementations. In general, DHTs store key-value pairs and provide a core *lookup* operation: given a key of some key-value pair, map the key efficiently to a node on which the value resides. When the key-value pair is a name-name pair (as with DNS), the DHT provides the name resolution infrastructure. This is the model advertised by LNA and that we shall describe here.[8] Data keys in Chord are flat m-bit binary strings (e.g. hashes of things) and those are uniformly mapped to available nodes in the infrastructure using consistent hashing techniques. The node ids (IP addresses) and the data keys are mapped into the same namespace of a cryptographic hash function. This hashing allows for a flat and flexible naming mechanism. Resolving an SID in LNA might simply be a key lookup in Chord which returns a value, the corresponding set of EIDs in this case. Chord is completely decentralized, and can lookup keys using only $\log(n)$ messages, where n here is the number of Chord nodes. Basically, in order to provide scalable lookup, Chord works as follows: the namespace of order $2^m - 1$ is logically arranged as a ring as shown in Fig. 3.4. A distance function is defined between any two names in the namespace to be their distance on the ring modulo 2^m. Each node acquires a name from the $2^m - 1$ namespace which is a hash of its IP address for example. The node with name p stores all the names q such that q is closer to p than to any other node's name in the DHT. Looking up a name r then is simply finding the node in the DHT whose name is closest to r under the distance function. In order to locate such node, each node in the DHT keeps pointers to a logarithmic number of other nodes at various distance scales. Chord is self-organizing and its lookup mechanism survives node failures and re-joins. The organization of the namespace according to the distance function is the main reason behind the scalability of DHTs and is at the same time troublesome in context of Internet naming as we shall discuss next.

3.5.2.1 Properties

On the positive side, the flat namespace of LNA is inherently more flexible than that of DNS. It promotes wider applicability as it is not constrained by the hierarchical domain structure and ownership and may as a result incorporate any type of object. Additionally, by their decentralized and self-organizing nature, DHTs are more resilient than hierarchical DNS-like systems, they require less aggregate cost to maintain and operate, and they gracefully evolve and scale with demand. There are, however, several challenges to a flat namespace with a DHT-based resolution system. First, humans are not good with flat names; this lack of semantics in SIDs requires yet another layer (and complexity) to map from human-readable names to SIDs. Second, and more importantly, there is no clear economic incentive to sharing the cost of the global DHT. While in the current DNS system, each party owns

[8]Other models exist where the key-value pairs correspond to the name-object pairs and the lookup provides some overlay forwarding mechanism to the objects.

and operates its own subspace and bears the cost, in DHTs it is unclear how the cost is to be shared among the participants especially given that the names owned by an organization might end up being stored on a DHT node managed by some other organization. This raises yet another issue of trust and authoritativeness over resolution which is unclear in a DHT solution. Lastly, with flat names more elaborate mechanisms are needed to guarantee name uniqueness [29]. While we have presented these challenges in the context of LNA, they are common to flat naming proposals such as the Host Identity Protocol (HIP) [118], and the Internet Indirection Infrastructure (I3) [149] and [121] to mention a few.

3.6 Conclusion

In this chapter, we have formalized the definitions of name, address, forwarding, and mapping, and we have distinguished routing from discovery. There are several key takeaways that the definitions tried to highlight:

- A name alone is generally of limited value; we are interested in the name only in relation to the object(s) it names.
- We need to distinguish the association of the name to the object it names from the association of two objects. Only the latter is referred to as the binding.
- Only after we specify the binding of an object to another object can we semantically distinguish the object's name from its address. The address is the name of the object to which it is bound.
- A name is relative and to define the scope within which it is valid we need to tie it to forwarding.

We presented two case studies of naming architectures with different (and somewhat contrasting) requirements and we discussed their properties and challenges. Our hope is that this chapter eliminates the terminology confusion, highlights the requirements and challenges in designing naming systems, and sets the stage for the coming chapters.

Chapter 4
A Taxonomy of Naming Architectures

Everything should be made as simple as possible, but not simpler
Albert Einstein

We saw in Chap. 2 that naming exists within the larger context of the network architecture which is intended to satisfy a set of user requirements. We surveyed a multitude of internetwork and naming architectures and proposals based on their high-level service model whether communication, information, and/or computation. We have seen that while the communication structure is necessary for defining and representing a network architecture, it is not sufficient. Information and computation structures are building blocks that need to be properly understood within modern network architectures. Chapter 3 then presented some simple definitions and abstractions. We defined network objects along with their binding relation, and names along with the naming relation. A name was defined relative to the object(s) it names, and its validity (or scope) was defined relative to the network forwarding function. This chapter bridges the gap between the high-level survey of Chap. 2 and the formal definitions of Chap. 3. Our goal is to devise a simple taxonomy that accommodates the multitude of naming architectures presented so far. In addition to framing the solution space, the taxonomy is intended to serve as a tutorial, to present a consistent terminology, to help find similarities and differences among architectural designs, and more importantly to help identify new avenues for contributing to the topic.

4.1 Introduction

As clearly highlighted by the survey of Chap. 2, there seems to be a growing consensus in the community about the need for designing a smarter network that is more than just a transparent "bit-plumbing" medium. While such evolution into a smarter and more complex Internet is bringing new potentials and service models, the community generally lacks frameworks (or taxonomies) for thinking about such models and their design implications. In fact, the majority of the recent architectural work is either aimed at exploring novel usage models that adhere to a class

J.S. Khoury, C.T. Abdallah, *Internet Naming and Discovery*,
Signals and Communication Technology, DOI 10.1007/978-1-4471-4552-3_4,

of applications, or at directly addressing a set of limitations of the current Internet. Content networking is an example of a usage model that is gaining traction. On the other hand, specific limitations with the current architecture, as discussed earlier in Sect. 1.3, include the lack of mobility support, the lack of security and trust, and the economic mismatches.

Our goal is to devise a simple taxonomy that accommodates the multitude of current naming architectures, and potential future ones. While trying to devise the taxonomy, it seemed to us that the body of network and naming architecture work in general is difficult to classify due to the independent nature of the many contributions to the field. However, we have noticed that modern networks are becoming increasingly intelligent, and the intelligence is being manifested by introducing more processing (e.g., [74]) and storage elements (e.g., [14]), and by providing the users with richer instruction sets instead of the simple static IP packet. Interestingly, such evolution resembles the computer architecture field, in the sense that the network architecture is being designed to provide a general-purpose platform to its diverse users. We shall leverage this analogy to directly apply some useful taxonomical notions from the computer architecture field to our work (e.g., [73, 82]).

We start by reviewing some basic definitions in Sect. 4.2. We then present the taxonomy in Sect. 4.3 before concluding.

4.2 Background

Before we present the taxonomy, let us first revisit the basic definitions from Chap. 3. We define Data Objects (DOs), belonging to universe $\mathcal{O}$, as network objects that participate in communication, of which Switching Objects (SOs) are a subset $\mathcal{S} \subseteq \mathcal{O}$. SOs additionally have internal switching logic and state which allows them to actively participate in message forwarding. We define the binding binary relation $B \subseteq \mathcal{O} \times \mathcal{O}$ between objects as the set of ordered pairs (o, o'), $o, o' \in \mathcal{O}$, meaning that object o is bound to object o'.

We define the universe of names $\mathcal{N} = \{0, 1\}^*$ to be arbitrary binary strings. We define the naming relation $N \subseteq \mathcal{O} \times \mathcal{N}$ as the set of ordered pairs (o, n) meaning object o is named by name n. We noted that under N an object may have multiple names and a name may be associated with multiple objects. Define the set of all objects named by name n, $objects(n)$, as follows:

$$objects(n) = \{o \in \mathcal{O} : (o, n) \in N\}$$

The forwarding relation, *forwards*, of quadruples is defined by $F \subseteq \mathcal{S} \times \mathcal{N} \times \mathcal{O} \times \mathcal{N}$ as a simplified abstraction of the switching logic and state of all SOs in the network. An ordered quadruple $(s, n, o, n') \in F$ where $(s, o) \in B$, $s \in \mathcal{S}$, $o \in \mathcal{O}$, and $n, n' \in \mathcal{N}$ means that when presented with some *message* with name n in its header (name of the intended target), SO s forwards the message to object o rewriting the name in the header to n'. The reachability function $reach(s, n)$ is defined as the set of objects that are reached by the forwarding function (the transitive closure of F)

when a message with name n arrives at SO s. The *scope* of a name n that names object o is defined relative to the forwarding relation as the set of all SOs that can reach o when presented with name n,

$$scope(n, o) = \{s \in \mathcal{S} : o \in reach(s, n)\}$$
$$scope(n) = \{s \in \mathcal{S} : reach(s, n) \neq \phi\}$$

Having reviewed the basic definitions, we present the taxonomy next.

4.3 Taxonomy

Our taxonomy isolates the core elements of a network architecture relevant to naming. We classify *instructions*, *addressing modes*, and *addressing types*. The taxonomy is descriptive in nature. It is concerned with low level naming primitives from which arbitrarily sophisticated naming services are generally composed. By focusing on the basic underlying primitives, the taxonomy stands a better chance at being general enough for classifying the multitude of current architectural proposals, and potential future ones.

4.3.1 Instructions

Switching Objects (SOs) represent protocol units and are the main processing elements within our simplified network model. The network SOs collectively execute instructions to deliver the desired service model. While every network architecture comprises different SOs and protocols, there are generally four classes of instructions within a general-purpose architecture: *Data Transfer*, *Data Manipulation*, *Control*, and *Special Purpose*. We define the instruction class in terms of how the instruction affects the message. A recent collaboration between academia and industry is in the process of defining an instruction set architecture for a general-purpose programmable switch (see the OpenFlow specification at www.openflow.org).

- Data Transfer: instructions transfer inbound messages to output ports or to the SOs internal memory. The IP packet forwarding is a good example here. An IP packet instructs the IP network layer SOs to transfer/deliver the packet to a destination SO, which is translated to an output port. The messages may be transferred to a memory location instead for in-network caching and storage (e.g., [5, 166]). The IP address of the destination is included in the packet header. The data and interest packets in [90] are examples of transfer instructions towards content.[1] Besides header rewrites, data transfer instructions do not otherwise affect the contents of the message.

[1]The architecture [90] advocates the host-to-content paradigm instead of the host-to-host paradigm of TCP/IP.

- Data Manipulation: instructions are transfer instructions that additionally manipulate the contents of the message. In-network content aggregation, reduction, transcoding, encryption, decryption etc. are examples of potential data manipulation instructions.
- Control instructions: are intended to help a data message be successfully processed, mainly forwarded, stored, and/or manipulated. Example instructions here include those for status updates (responses, ACKS, NACKS, etc.) and statistics, and management of internal naming, forwarding, and other protocol state.

4.3.2 Naming/Addressing Modes

At the lowest level, a switching object (SO) forwards messages from input ports to output ports. Each message has a header containing a set of names. Naming/addressing modes define how these names identify "where to send the message". Addressing modes are defined relative to the SO. We distinguish four different addressing modes: *port*, *pointer-direct*, *pointer-indirect*, and *associative*. A port address directly identifies the output port(s) to use. A direct pointer is mapped internally by the SO into a valid port(s) onto which to forward the message. Indirect pointers are instead mapped through external mechanisms into valid direct pointers and/or port(s). A SO keeps internal state about direct pointers in order to correctly map them to ports. Associative names are descriptions of the object which are mapped into direct or indirect pointers and or ports.

- Port: source routes are a good example of port addressing whereby a message contains port addresses that a set of SOs along the message's path utilize for direct forwarding. If the source SO s knows a-priori the full path that a message needs to traverse, say the path is $s \rightarrow s_1 \cdots \rightarrow s_t$, then the source includes the full path—the source route ($s \rightarrow s_1 \cdots \rightarrow s_t$)—in the message header. Each SO along the path interprets the source route in the header, determines the next hop SO and the corresponding port, and forwards the message on its way. In layered architectures, source routes generally operate within a layer i.e. within a subset of the SOs. Instead of delegating the path discovery to the network completely, port addressing provides fine grained control over the path(s) that a message traverses. Port addressing allows for the most efficient form of routing but requires a-priori route discovery. It also provides several other advantages. We refer the reader to [165] for more on the advantages of user selected routes for the Internet.
- Pointer-Direct: This is the most common mode of addressing used by link-state routing protocols. It allows for efficient message routing. A SO keeps an internal table/map from pointer addresses to output ports. Upon receiving a message, the SO locates the pointer address in the message header, performs a fast lookup on the address to obtain the corresponding output port(s) from its internal table, and switches the message. The Internet IP address is a pointer-direct name within the IP layer SO. The hierarchical nature of the IP address allows the routing system to use aggressive aggregation for efficiency. The hierarchical persistent content

names in [16, 90] and flat names in [42] are pointer-direct names as well within the network layer SO. For efficient routing, aggressive aggregation is performed on the hierarchical names of [16, 90] while smart DHT-based routing is employed in [42].

- Pointer-Indirect: SOs map pointer-indirect addresses through external mechanisms into valid direct pointers and/or port(s). This is generally a costly operation during path discovery. Indirect addressing, however, allows for cheaper updates when name mappings change as in the case of dynamic networks. DNS names, described earlier in Sect. 3.5.1 are one form of pointer-indirect addressing. A pointer-indirect DNS name is mapped into the pointer-direct IP address, which is then mapped to outbound port(s) for forwarding. Identifier-locator split proposals utilize this mode of addressing with flat endpoint identifiers (EID) (see [29, 66, 118, 119, 158]).
- Associative (or intentional addressing): this form of addressing is used when the name of the target object is unknown but some properties and/or attributes about the object are known. The message header contains an intentional "description" the object which is eventually mapped to one or more ports using some internal or external mechanism. Example architectures and proposals that employ associative addressing include [20, 54, 161].

4.3.3 Naming/Addressing Types

Naming types are classified into *unicast*, *k-cast*, and *multicast*. The type of a name n is determined by the set of network objects addressed by n. When n addresses a single object, it is a termed a unicast name. When it addresses more than one object, say m objects, it is termed multicast or k-cast. Multicast names address the whole group of m objects, while k-cast names address any k of the m objects. Anycast [128] is an alternative term for 1-cast ($k = 1$).

Formally, we define the types as follows.

Definition 4.1 Multicast (and Unicast): When $reach(s, n) = objects(n)$, $\forall s \in scope(n)$, then name n names the set $objects(n)$. If $objects(n) = \{o\}$ is a singleton, then n is a unicast name. If instead $|objects(n)| = m > 1$, then n is a multicast name. Finally when $|objects(n)| = |\mathcal{O}|$, then n is a broadcast name.

In the multicast (and broadcast) definition we are assuming that $objects(n)$ is the group of objects named by the multicast name and hence $\exists (o, n) \in N, \forall o \in objects(n)$ i.e. an ordered pair $(o, n) \in N$ exists for each named object in the group.

Definition 4.2 k-cast (and Anycast): When $|reach(s, n)| = k < |objects(n)| = m$, $\forall s \in scope(n)$, then name n is a k-cast name i.e. it names any k of the m objects $objects(n)$. For the special case when $k = 1$, then n is termed an anycast name.

In the same way as multicast, we are assuming that an ordered pair $(o, n) \in N$ exists for each named object in the group *objects*(n).

Having defined the types, we give some examples next.

- Unicast: Almost all network architectures provide a unicast primitive. Unicast has been the fundamental abstraction and communication primitive within addressing and routing design. Whether unicast or multicast (discussed next) should be the basic primitive is an interesting question given that physical communication media are inherently multicast [127].
- Multicast: is important for efficient communication among a group of objects. When the architecture does not directly provide a multicast primitive, the latter may be implemented as repeated unicast. Repeated unicast is, however, inefficient and several modern proposals explicitly provide a native multicast primitive (e.g., [16, 77, 90, 149]). The Internet TCP/IP does not provide an internetwork IP multicast primitive. Several proposals have been developed for network layer reliable multicast (e.g., PGM RFC 3208), and best-effort multicast (e.g., IGMP RFC 3376).
- k-cast (and Anycast): the anycast [128] (and k-cast) primitive is particularly useful for efficient load-balancing. A service is load-balanced by being replicated across several network objects all sharing the same anycast (or k-cast) name. Users seeking the service using the anycast (or k-cast) name are directed to the closest (or k-closest) available copies.[2] Anycast has been proposed as an efficient primitive for content distribution and service load-balancing in content-centric networking (e.g., [16, 90, 99]). We are not aware of any proposals with a native k-cast primitive for $k > 1$.

The unicast, multicast, anycast, and broadcast primitives enable 1-to-1 and 1-to-M addressing and forwarding. A more complex M-to-N communication service may be composed out of these basic primitives. For example, an interest packet in a publish-subscribe architecture (such as [90]) may be forwarded to M sources (publishers) who then publish the data to the N targets (subscribers).

4.4 Discussion and Conclusion

This chapter presented a simple taxonomy of network naming and addressing architectures. We classified architectures based on three low level primitives: instructions, naming/addressing modes, and naming/addressing types. Arbitrarily sophisticated naming services are generally composed out of these primitives, as we saw in the survey of Chap. 2. We decided to focus only on the basic underlying primitives in order to keep the taxonomy general enough to be applicable to the multitude of current architectural proposals, and potential future ones. In terms of its value, the taxonomy serves as a tutorial for introduction to the field. It additionally helps

[2] The object selection does not need to be based on closeness in some metric space, however.

in organizing and thinking about the architectural space, in finding similarities and differences among architectural designs, and more importantly in identifying new avenues for contributing to the topic. Finally, in terms of its limitations, the chapter in its current form falls short of providing tangible outcomes beyond the descriptive one. This chapter concludes our discussion on naming architectures. We started Chap. 3 by saying that names ultimately help us identify and discover objects on the network. The next two chapters examine name-based discovery.

Chapter 5
On the Feasibility of Dynamic, Efficient, and Scalable Name-Based Discovery Systems

Not all those who wander are lost
J.R.R. Tolkien

Chapter 3 formalized the definitions of name, address, locator, binding, discovery, and routing. We have defined name-based discovery in Sect. 3.4 as the end-to-end process by which a message reaches its intended destination object starting with a name. We made an explicit distinction between *discovery* and *routing* by stating that routing is the subprocess that acts directly on the main message while discovery additionally includes other mechanisms that assist in correctly forwarding the message without directly acting on it. We gave examples of such additional mechanisms including DNS and ARP name mapping. Accordingly, we claimed that discovery, the process of discovering a path to the destination(s), subsumes routing. We shall use the general term discovery throughout, and make the distinction between discovery and routing only as needed.

Traditionally, the design of name-based discovery systems on the Internet has assumed the existence of some underlying "efficient" routing scheme(s) that is location-aware along with a means to map from higher level names to location-based routing identifiers (or locators). The alternative is to redesign the discovery scheme to operate on location-independent names (or identifiers) without the need for the locators. The literature is replete with such proposals, and indeed, there has been renewed attention in the networking community on revisiting the question: what if we start with persistent names and attempt to design a discovery scheme whose main currency is the persistent names? Starting with persistent names is appealing as it eliminates the notion of "location" along with the complications that come with overloading (or even separating) identity and location identifiers. This potentially allows named objects to relocate more seamlessly. The notion of discovery/routing on persistent names, however, raises many questions on the feasibility of such approaches. This chapter explores general models aimed at studying such alternatives for designing name-based discovery systems for the Internet.

Designing a discovery system is commonly challenged with providing three desirable properties: *scalability*, *efficiency*, and *adaptability*, which we describe shortly. While there are several other design dimensions involved, we focus on these

J.S. Khoury, C.T. Abdallah, *Internet Naming and Discovery*,
Signals and Communication Technology, DOI 10.1007/978-1-4471-4552-3_5,

three to keep the design space tractable and to explore some fundamental tradeoffs a designer is generally faced with. We start by reviewing the benefits of models, and we proceed to explore two models for discovery in computer networks. The first is a general model that nicely highlights the tradeoffs across a wide range of networks, while the second model is geared towards complex networks exploiting the *small-world* structural property. We show how models help us better understand and reason about the design tradeoffs.

5.1 Introduction

Models help us understand the world around us. With models at our disposal, we can better decide, strategize, and design. More importantly, models help us derive results that would not otherwise be intuitive or predictable. To mention a few examples, supply–demand curves help us predict the impact of demand changes on prices. The carbon cycle model helps governments identify where to intervene to have the most policy impact. Risk models help federal governments decide which companies to bail out post an economic crisis. The mechanism design framework helps institutions design mechanisms to achieve better social outcomes. Game-theoretic models help us predict equilibria for interactions among strategic agents.

In order to devise a general-purpose and realistic model of discovery in networks, we need to understand the parameters of interest to the model. To do so, we start by defining some basic terminology. For the purposes of this chapter, we represent a communication network with an undirected graph $G = (V, E)$ where V is the set of nodes and E is the set of edges, $|V| = n$. We say that nodes u and v are neighbors if there exists an edge $(u, v) \in E$, and we denote by $B(u)$ the set of direct neighbors of u i.e. $B(u) = \{v : (u, v) \in E\}$. We assume each node u is identified with a unique name $name(u) \in \mathcal{N}$. Given the name $name(v)$ of some destination node v, our goal is to devise a name-based discovery scheme that can deliver any message $M(name(u), name(v))$ (originating at source node u) to its intended destination v. The message contains the name of the destination in its header. Finally, we assume each node w uses a local *forwarding* function f_w, where $f : \mathcal{N} \rightarrow E$ maps from a name to an incident edge $(w, z), z \in B(w)$ on which to forward the message. Node w additionally assigns to each incident edge (w, z) a locally unique label denoted by $l(w, z)$ such that $l(w, z) \neq l(w, x), \forall z \neq x$. Note that, following the definitions of Chap. 3, we use the term *name* generically and we explicitly specify the object being named as well as any assumptions we make regarding the properties of the name (e.g. flat, topological, etc.). Our intentions should be clear from the context.

Several questions come to mind as we first start working through a model. Does the designer of the discovery scheme get to choose the node names or are those assigned arbitrarily by some external party? How do we choose the names? What are we assuming about the structure of the graph G? and so on. Let us revisit the simple ARPAnet design principles, introduced earlier in Chap. 1, and work through

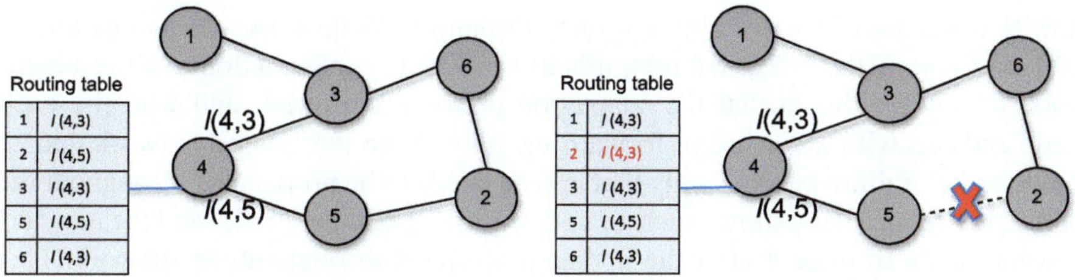

(a) Before update (b) After update

Fig. 5.1 Example ARPAnet IMP network. Node 4's routing table is shown (**a**) before the failure of link (5, 2), and (**b**) after link failure and protocol adaptation

a simplified design of the original IMP-to-IMP[1] network numbering and routing scheme as an example. Recall that a core design principle of the APRAnet was reliability in the presence of link and node failures, hence the network was assumed to be dynamic. Additionally, the first IMP network was simple and comprised 19 nodes with redundant links, and no limiting assumptions were made regarding its structure. We show an example $N = 6$ node IMP network in Fig. 5.1. We assume node names are arbitrarily chosen from the integer range $[1, 6]$. The name-based discovery scheme shown in Fig. 5.1 works as follows: each node maintains internal state about the rest of the network in the form of a forwarding table. The forwarding table of node v is an indexed table $(name(w), l(v, u)), \forall w \in V$ where $u \in B(v)$. Figure 5.1 shows the forwarding table of node 4. When presented with a message $M(name(u), name(z))$ destined to node z, the forwarding function of node v simply locates the entry corresponding to $name(z)$ in its local forwarding table, determines the label of the edge on which to forward the message, and forwards M accordingly. For example, in Fig. 5.1(a), node 4 knows to forward message $M(., 2)$ on link $(4, 5)$ by looking up entry $(2, l(4, 5))$ in its forwarding table. Clearly, each node knows its name and can tell if the message has reached its destination by comparing the destination name to its own. A control protocol helps each node compute its local forwarding table such that the tables are consistent. One such adaptation control protocol that is intended to route messages on shortest paths (in terms of number of hops) works as follows: at each iteration, a node computes a best estimate of its distance (in number of hops) to each other node and exchanges this reachability information with all its neighbors. A node that receives reachability updates from a neighbor node (i) computes its reachability information to the rest of the network in terms of shortest paths, (ii) updates its local forwarding table accordingly, and (iii) forwards its reachability information to its neighbors (only in case of changes). For example, node 4 in Fig. 5.1(a) receives reachability updates from its neighboring nodes 3 and 5. Node 5's update contains an entry $(name(2), 1)$ indicating that 5 may reach node 2 in one hop, while node 3's update contains an entry $(name(2), 2)$ as node 3 needs two hops at least to reach node 2. Accordingly, node 4 now knows

[1]Interface Message Processor (IMP) is analogous to the router.

how to reach node 2 in two hops by going through 5. Node 4 updates its forwarding table to $(name(2), l(4,5))$ and forwards its reachability information to all its neighbors. It may be shown that the adaptation protocol converges and whereby each node ends up with a consistent forwarding table. Note that after a network update (node or link failure or join), only that update needs to be propagated throughout the network. Figure 5.1(b) shows such an update (failure of edge $(5,2)$) and the new forwarding table of node 4 after the update propagates throughout the network. This adaptation protocol is an extremely simplistic version of the Border Gateway Protocol (BGP) [132] that controls inter-domain routing on the Internet today (nodes are internet routers in the Default-Free Zone (DFZ)). BGP in reality is a policy-based path-vector protocol that chooses best routes based on operator policies and preferences over routes [67].

The discovery scheme we just explained seems pretty powerful as it deals with dynamic networks, with arbitrary network structures, and with node names chosen arbitrarily. The scheme is able to route messages "efficiently" along shortest paths. Let us, however, examine its shortcomings. First, if we define the *degree* of a node to be the number of edges incident to the node, then a node with degree d must store a forwarding table of size $N \log d$ bits. More clearly, we have N labels $l(u,v)$ in the forwarding table, and each label is represented with $\log d$ bits. Note that the names here are simply the table indices, so no need to explicitly store them. Hence, the size of the table, the *state*, grows linearly with the size of the network N and the scheme is accordingly not "scalable". Indeed this BGP routing table scalability problem on the Internet has received significant attention in the past decade [88, 115]. The Internet routing table grows exponentially due to address space de-aggregation and provider-independent addressing, and traffic engineering practices. To better appreciate why scalability is important, we note that the cost of a DFZ router grows linearly with the size of the routing table it must support [83]. This network externality affects the bottom line of network operators who must continuously invest in new routers as the network grows. Second, each topological change must be advertised globally requiring at least $O(|E|)$ messages to be exchanged before the protocol reaches a consistent state (it is even worse than that in BGP). In other words, the scheme has *adaptation cost* linear in the size of the network. The adaptation cost has direct implications on the convergence times, and the stability of the protocol in general [76]. The larger the protocol's adaptation cost, the higher the actual operational cost of the network which directly affects operators' bottom lines.

The example above is insightful as it highlights the key design tradeoffs, mainly efficiency, scalability, and adaptation cost. A model helps us understand the limits on such tradeoffs so we can determine whether it is possible to do better than linear scaling for example, and design accordingly. In addition, the example highlights several parameters that a good model needs to capture. Are the node names chosen arbitrarily or can the designer embed location information into the names to potentially perform better? Is the graph static or dynamic? Does the scheme assume specific graph topologies (e.g. trees), or universal graphs? We shall introduce the first model, compact routing, in the next section and distinguish its many variants: namely labeled vs. name-independent, static vs. dynamic, and universal. For each of the variants, we review the main results. The results are typically stated in terms of tradeoffs and bounds on efficiency, scalability, and adaptation cost. The second model is then introduced in Sect. 5.3.

5.2 Compact Routing Model

5.2.1 Background

We introduce the compact routing problem here, which we shall refer to throughout the rest of the book, and we review its two variants: *labeled* versus *name-independent* compact routing. We refer the reader to [69, 71] for surveys on the topic. We reuse the same terminology introduced earlier and we additionally assume that the undirected graph $G = (V, E, w)$ is weighted. An adaptive, compact routing scheme routes messages between nodes with the goal of optimizing *stretch*, *space*, and *adaptation cost* defined as follows: Let $p(u_1, u_k)$ denote a simple path between nodes u_1 and u_k, and let its length be $|p(u_1, u_k)| = \sum_{i=1}^{k-1} w(u_i, u_{i+1})$, the sum of the weights of all edges on the path. The stretch of a path $p(u, v)$ is defined as the ratio $\frac{|p(u,v)|}{d^*(u,v)}$ where $d^*(u, v)$ is the length of the shortest possible path between u and v i.e. $d^*(u, v) = \min_{p(u,v)}(|p(u, v)|)$. The stretch of the routing scheme, which finds paths $p(u, v)$, is defined as the maximum such ratio among all source-destination pairs in G i.e. stretch $= \max(\frac{|p(u,v)|}{d^*(u,v)}), \forall u, v \in V$. Space is the size of the routing state (in bits) stored per node. Adaptation cost is the amortized cost in terms of message and time complexity needed for the routing algorithm to converge in response to a topology change, such as node addition or deletion.

Over the past three decades, the theoretical computer science community has focused on developing the theoretical limits and tradeoffs between stretch, space, and adaptation cost. Several variants of the problem have been studied mainly *static* vs. *dynamic*, *labeled* vs. *name-independent*, *universal* vs. *topology-dependent*, and *single-source* vs. *any-pair*. The static compact routing problem assumes a fixed topology and is accordingly not concerned with adaptation cost. Labeled (or name-dependent) compact routing (LCR) allows the designer of the routing scheme to choose node names to better suit the routing scheme, giving her more control by potentially embedding topological information into the name. On the other hand, name-independent compact routing (NICR), first distinguished in [27], assigns to nodes arbitrary unique names. As the name suggests, universal schemes make no assumptions about the graph topology. Finally, compact routing distinguishes between *single-source* routing where the scheme is only concerned with routing from a designated source node to all destinations (generally performed on a tree with source = root), or *any-pair* routing where any node should be able to route to any other node.

For static routing schemes, there is clearly a tradeoff between space and stretch: the more information nodes keep about the graph (hence the more space), the better paths they can choose for routing (hence smaller stretch). For example, as we

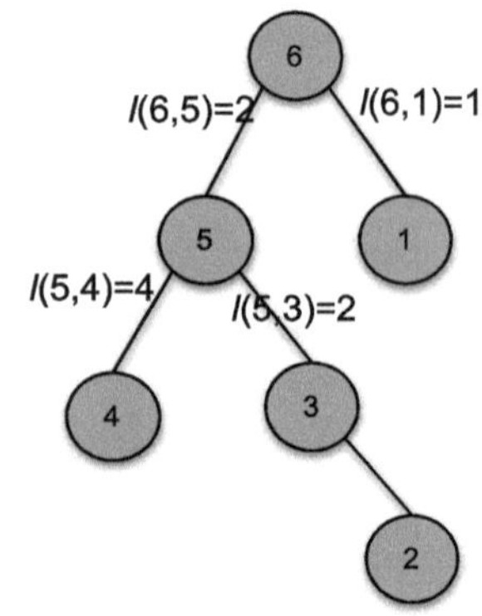

Fig. 5.2 Sketch of LCR scheme on trees. Node names are distinct integers chosen from the interval $[1, n]$ as the DFS numbers

showed in the introduction, a trivial shortest path (stretch 1) scheme may be devised when each node keeps state about every other node in the network. More precisely, if each node u with degree-d keeps the next-hop interface id of each destination on a minimum spanning tree rooted at u, the per-node space requirement is $O(n \log d)$. This assumes nodes are identified with integers in the range $[1 \ldots n]$ and hence a node only maintains an indexed table of next-hop interface ids, alternatively referred to as the routing table. A natural question is whether we can do better than $O(n \log d)$ bits per node for shortest path on universal graphs. Before answering this question for universal graphs, let us examine the simpler case of static acyclic graphs (trees).

Santoro et al. [139] provide an optimal labeled routing scheme on trees. The scheme belongs to the class of Interval Routing (ITR) schemes since each node assigns intervals to each of its incident links on which to forward messages. The scheme works as follows: starting from some node $u \in V$, compute a minimum spanning tree T of G rooted at u. Assign the node names as follows: traverse the tree in depth-first style (DFS) and assign to each node a distinct integer (called the DFS number) when it is reached for the first time; the integers are assigned in decreasing order. Figure 5.2 shows a sketch of an optimally labeled tree with node names $name(u)$ chosen as the DFS numbers. Each node u additionally assigns a label $l(u, v) = \min(name(w) \mid w \in T[v])$ to each $v \in B(u) \backslash \{parent(u)\}$, where $T[v]$ is the subtree of T rooted at v. We show these labels as assigned by nodes 5 and 6 in Fig. 5.2. The optimal labeled routing scheme works as follows: when presented with a message $M(., d)$ destined to d, node w checks first if $d = name(w)$. If true then the message has reached its final destination, otherwise if $d > name(w)$ or $d < \min_z(l(w, z))$, then forward M on $(w, parent(w))$, else find $\max_z(l(w, z)) : l(w, z) \leq d < l(w, z')$ and forward M on (w, z). For example, node 5 in Fig. 5.2 forwards $M(., 3)$ on $(5, 3)$ since $l(5, 3) = 2 \leq 3 < l(5, 4) = 4$. It may be easily shown that the algorithm is optimal i.e. routes on shortest paths, and each node maintains $O(d \log n)$ bits of state (its name and the labels) which is logarithmic in the size of the network. We quickly note here that if we allow the locator (alternatively message header) size to increase, the space can be further optimized

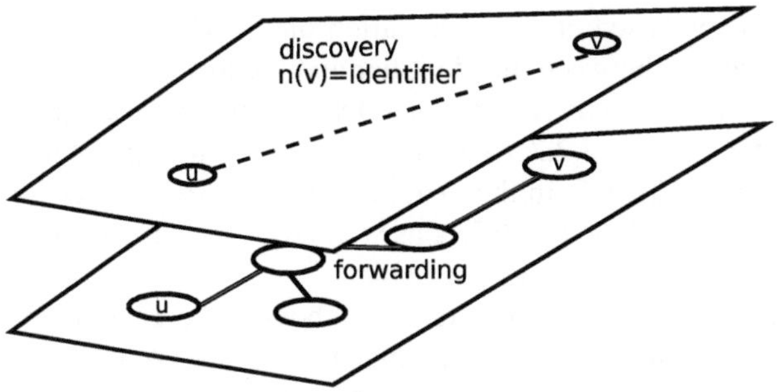

Fig. 5.3 Decoupled naming and forwarding layers. Upper layer performs discovery on location-independent names (or identifiers) while lower layer forwards using location dependent names (or locators)

to $O(\log n)$ bits instead (see [64, 157]). Notice how the names are chosen to follow the topology making its easier for the routing scheme to discover optimal paths.

Having seen an optimal LCR scheme on static trees, we revisit the question of whether we can we do better than $O(n \log d)$ bits per node for shortest path on universal graphs (each node keeps state about the whole network)? Unfortunately, the answer is negative as shown in [72]. Briefly, there exists a graph topology that requires $\Theta(n \log d)$ bits per node for optimal stretch-1 routing and hence any universal shortest path labeled compact routing scheme requires $\Omega(n \log d)$ bits per node (for a total of $\Omega(n^2 \log d)$ bits on all nodes). What this means is that in order to obtain a better space bound, we must trade off stretch (stretch must increase). For universal graphs, Gavoille et al. [70] show that in order to achieve a stretch factor strictly less than 3, each node must keep $\Omega(n)$ bits for a total of $\Omega(n^2)$, which is the best that any universal routing scheme may achieve (necessary condition). Note that an infinitely scaling universal LCR scheme of stretch factor 3 has been presented by Thorup and Zwick and shown to require $O(\sqrt{n})$ space [157]. The results so far have assumed labeled compact routing with header relabeling. As we saw, labeled compact routing schemes choose names that are topology dependent. For example, in the Santoro scheme we just reviewed, a node name depends on its position in the DFS search.

On the other hand, name-independent compact routing chooses names arbitrarily (names are independent of location). While we have used the term *name* generically so far, NICR explicitly distinguishes a location-independent name. More generally, this is a type of persistent name that is independent of the attributes of the object being named. In the compact routing model, the object being named is the node, and the attribute of interest is node's location. As introduced in Chap. 3, this type of persistent name is referred to as the *identifier*, while a location-dependent name is referred to as the *locator*. In a sense, NICR decouples the naming and the message forwarding functions into separate layers as shown in Fig. 5.3. The upper layer performs discovery on location-independent names (or identifiers) while the lower layer forwarding function (generally) utilizes location dependent names (or locators). Intuitively, one would expect the stretch for NICR to be at best larger (or equal) than that of LCR. This is based on the observation that a NICR scheme has to discover the additional binding between the identifier and the locator(s) during the discovery phase, a step which generally increases stretch. A concrete example of this layering is presented later in Sect. 5.2.2 where the lower forwarding function uses optimal LCR. As we saw in Chap. 2, in the extreme case, the lower layer forwarding function could operate solely on identifiers as well (see [16, 42] for examples). Such an approach, however, tends to be less scalable.

When trading off stretch for a better space bound, the general approach used by compact routing schemes is to split the namespace into compact sets of sub-linear size, for example $\sqrt{n}$. Each of these sets forms a group (or a color) and state about each group is distributed throughout the network. A node keeps state about a subset of the groups (instead of the whole network). Looking up a name then requires first identifying the group to which the name belongs, then contacting a nearby group representative who generally knows about the locators of all names in the group. For example, in the stretch-3 single-source routing scheme on trees due to Arias et al. [26], any graph degenerates into a spanning tree rooted at the source. Each of the neighboring nodes of the root node maintains state about a distinct group allowing the root to lookup any destination node by referring to the respective group representative node which is one of its neighbors. In the same sense, in any-pair routing on universal graphs, each node knows about all the group representatives and those generally belong to the node's neighborhood. For example in the optimal stretch-3 routing scheme by Abraham et al. [19] on general graphs which we shall elaborate on shortly (and the previous stretch-5 by Arias et al. [26]), this is exactly the case. Each neighborhood is *fully colored* and a node that represents the group/color knows about all nodes in the group. The challenge with such schemes lies in the means to distribute the group responsibilities to nodes (i.e. which node represents what groups?) such that (1) each neighborhood has at least a representative for each group, and (2) no node represents more than a logarithmic number of groups. It has been shown in [26, Lemma 3.1] that such assignment exists and is computable in polynomial time. Next we present a concrete example to illustrate optimal universal NICR on static graphs and show how the scheme may be extended to name arbitrary objects in addition to the graph nodes.

5.2.2 *Optimal Universal NICR*

Abraham et al. [19] developed a universal NICR scheme that is optimal in the strict sense, i.e., requires $O(\sqrt{n})$ space with stretch 3. Recall that this is proven to be the lower bound achievable by any compact routing scheme. This result is interesting as it proves that choosing arbitrary node names does not necessarily degrade the performance (scalability and efficiency) of the routing scheme. This is a good example of how models help us derive results that are otherwise unintuitive.

Briefly, the scheme operates as follows on a graph $G = (V, E, w)$: each node u keeps track of its vicinity ball $B_k(u)$ that includes the set of k closest nodes to

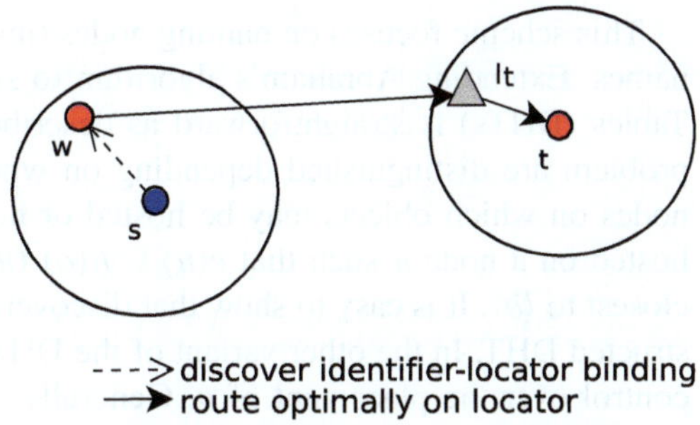

Fig. 5.4 Sketch of Abraham's name-independent compact routing scheme

u in G. Note that "closeness" is based on distance measure $d(u, v)$ which is the cost of a path from u to v, i.e., sum of weights on edges. The value of k is picked to be $8\sqrt{n}\log n$. Each node u has a color, $c(u)$, assigned from a set of $\sqrt{n}$ colors. The node's color is determined by hashing its name (identifier), and picking the first $\log\sqrt{n}$ bits from the hash $h(u)$ as the color. The number of nodes belonging to a particular color set is shown to be $\leq 2\sqrt{n}$. Any one of the color sets is chosen to be the landmark set L. Based on the balancing assumption which results from hashing, each node u will have a landmark node l_u in its vicinity. In Fig. 5.4, we show the vicinities of two nodes: a source node s with blue color (i.e. $c(u) = blue$) and a red destination node t as well as t's landmark node l_t. The routing scheme builds on optimal labeled compact routing in trees, which is shown to be performed optimally using $O(\log^2 n/\log\log n)$ space per node [64]. Hence, in terms of routing state, each node u maintains the following: (1) for each landmark node l, u's locator for the minimum spanning tree (MST) rooted at l requiring $O(\sqrt{n})$ space (note that routing on those trees is optimal stretch-1; (2) for each node $v \in B(u)$, node u's locator in the MST rooted at v requiring $O(\sqrt{n})$ space; and (3) for each node v having the same color as u, i.e., $h(v) = c(u)$, l_v's and v's locators in the MST rooted at l_v requiring $O(\sqrt{n})$ space but no additional MSTs.[2]

Given this $O(\sqrt{n})$ space at each node, it can be shown that routing is optimal with stretch-3 requiring message header re-writes [19]. Figure 5.4 shows how node s routes optimally towards node t, which has a different color just by knowing t's name (identifier) denoted as $\langle t\rangle$. Note that identifiers for the nodes are globally unique and are chosen arbitrarily from the integer set $\{1, 2, \ldots, n\}$. The identifier of a node is independent of topology, whereas its locator in some MST is topology dependent. Starting with the identifier $\langle t\rangle$, discovery involves two steps as follows: find a node w in my vicinity that has the same color as t (dotted arrow in Fig. 5.4). Node w is guaranteed to have a binding between t's identifier and a locator (the locator is a label in the MST rooted at l_t). Step 2 (solid arrows in Fig. 5.4) involves optimally routing on the MST. The stretch factor 3 follows directly from the triangle inequality.

[2]Note that the size of a neighborhood $B(u)$ is $4\alpha\sqrt{n}\log(n)$ which is very large, i.e., for $n \leq 65{,}000$, $|B(u)| > n$, and hence n has to be very large or otherwise each node will have to know about every other node. Hence, the Abraham scheme is not very useful for AS level graph where $n \approx 10{,}000$.

This scheme focuses on naming nodes (instead of arbitrary objects) with unique names. Extending Abraham's algorithm to support objects as in Distributed Hash Tables (DHTs) is straightforward as described in [18]. Two variants of the DHT problem are distinguished depending on whether the designer gets to choose the nodes on which objects may be hosted or not. In the first variant,[3] an object o is hosted on a node u such that $c(u) = h(o)$ (u has the same color as o), and $\langle u \rangle$ is closest to $\langle o \rangle$. It is easy to show that discovering any object o is optimal in the constructed DHT. In the other variant of the DHT problem, the designer does not have control over the placement of o. Generally, such model is employed for *locality-aware* closest copy routing. The concept of *locality-aware* means that the cost of locating an object o is proportional to the distance to the closest copy of o. Again, it is shown in [18] that the Abraham routing scheme may be extended incurring $O(\sqrt{n})$ pointers in the DHT per object name and maintaining scalability.

5.2.3 The Internet

How Internet Routing Works? The original design of the Internet routing system assumes aggressive aggregation of the addressing space where locators follow the hierarchical topology. This allows the routing system to scale as long as the topology and the addressing structure closely follow each other. Recently, however, provider-independent addressing, multi-homing, and traffic engineering practices have put strains on the routing system [115]. Multi-homing for example (a customer connecting to multiple providers), requires that a customer AS advertise a provider supplied prefix through its multiple providers. For example, assume the provider's prefix is 192.0.0.0/8 and part of it is delegated to the customer, say 192.10.0.0/16. This addressing structure follows the hierarchical topology (customer-provider) [67]. In the event that the customer connects to another provider for reasons of redundancy, the customer now advertises the prefix 192.10.0.0/16 through the other provider. This requires the first provider, who originally aggregated the customer's prefix, to de-aggregate the general prefix 192.0.0.0/8 and advertise the sub-prefix 192.10.0.0/16 as well or otherwise the more specific route through the second provider will always be used to reach the customer. The outcome is that the global routing table will now contain the two prefixes 192.0.0.0/16 and 192.0.0.0/8 due to de-aggregation instead of one. Provider-Independent (PI) prefixes as well may not be aggregated and each prefix requires $O(n)$ state in the global routing table (the BGP Routing Information Base (RIB) [132]) since every router in the Default Free Zone (DFZ) must keep state about the prefix. This deviation from the original design causes serious scalability issues with the routing system which again may only scale with aggressive aggregation. The reality is exacerbated by the fact that the number of BGP prefixes in the global routing table/RIB

---

[3] This is similar to structured peer-to-peer networks of single copy objects (e.g., [136, 150]).

is increasing exponentially at a rate of roughly 100,000 entries every 2 years and is expected to reach a total of 388,000 entries in 2011 [88]. Remedies to the scalability problem, such as [79, 152], at best scale linearly. Hence, while the original design of the routing system is scalable, the current reality is different.

In terms of efficiency, the Border Gateway Protocol (BGP), which controls Internet routes, is a policy-based path-vector protocol and is the de-facto protocol for Internet inter-domain routing. The protocol's specification [132] was initially intended to empower domains with control over *route selection* (which path or route to choose among multiple advertised routes to a destination), and *route propagation* (who to export the route to among an AS's direct neighbors) [43]. Route reachability information is broadcasted in BGP and nodes pick the routes that they value most which are not necessarily the shortest routes. Routing is intended to allow for a rich set of AS policies to be implemented [67, 75]. For example, if all ASes agree to implement shortest path, then BGP allows for it (but that is not the goal neither the reality as we shall explain in Chap. 8). We shall show later how hierarchical aggregation schemes (such as BGP [132]) are not optimal when it comes to the Internet topology.

Finally, in terms of dynamism we distinguish topology dynamics versus entity dynamics. BGP is designed to handle topology dynamics gracefully and to route around link or node failures. However, any such failure requires $O(n)$ communication which does not scale. For example, a link failure results in a BGP route withdraw message(s) that is propagated to all the DFZ. BGP dynamics are a major concern to the scalability of the Internet routing system [115]. In terms of entity dynamics, overloading of the entity name and the location name with the same IP address hinders mobility and portability [47]. If an entity wishes to move to a new location while keeping its name, then the permanent name of the new location must be changed which is very costly. This is perhaps the main drawback of using the same name for both location and entity objects. To remedy the problem, one of the recurring themes in the community is the need to separate the entity's identifier from its locator to enhance mobility (entity can move while maintaining the identifier) and security (trust information may be associated with the object at all levels). Check for example [47, 112, 118, 124, 129, 147, 149] for incremental proposals, and [29, 50, 120] for architectural approaches and considerations in this vein.

Can Compact Routing Help? The compact routing schemes discussed so far consider a static graph on which a data structure is constructed, ignoring the construction algorithm itself and the cost of constructing the data structures. As we have seen, the LCR schemes (and the NICR scheme that build on top of them) achieve their efficiency by choosing the node/labels names to closely follow the topology. This coupling between the labels and the topology, however, means that a topology change, such as the addition or deletion of an edge or a node, might require expensive changes to a subset of the node labels in order to keep the labels aligned with topology and retain the efficiency of the scheme. This renders static compact routing schemes impractical in dynamic settings where the structure of the network evolves over time.

Let us briefly overview how static routing schemes may be extended to support dynamics. Afek et al. [21] extend Santoro's static ITR scheme on trees described earlier (Fig. 5.2) to support growing trees where only leaf nodes can join. Afek's scheme requires a new DFS with a new index to be initiated each time a leaf node joins, resulting in larger node labels. A node's label depends on the DFSs in which it participated. The scheme maintains a stretch factor of 1 but requires $O(\log n)$ amortized message cost (number of control messages required post a topology change), and an increased label size of $O(d \log^3 n)$. An optimal and more dynamic scheme for compact routing on trees, which allows any nodes to join as well as nodes of degree 1 and 2 to leave the tree, is due to Korman et al. [100] and requires $O(\log^2 n)$ amortized message complexity and $O(\log n)$ label sizes. When extending an algorithm to support dynamics, it is not immediately clear what is being traded off in order to make the scheme adaptable. Afek et al. [21] showed that in order to obtain a constant stretch factor $<k$, the amortized adaptation cost (message and time complexity) of any routing scheme for universal dynamic networks of size n must be at least $\frac{n}{16k}$. This is a lower bound on the communication cost for universal graphs. In other words, no routing algorithm can scale better than $\Omega(n)$ in terms of adaptation cost on universal graphs. The validity of this result is based on worst case graph structures that force such a bound. But how about non-universal schemes over graphs with more structure, particularly, what about the Internet? We started the chapter by reviewing how the Internet routing scheme works in the DFZ. We showed how it scales poorly at it results in linear space growth, and worse than linear adaptation cost per topology change. A desirable discovery scheme for the Internet needs to be adaptable, efficient, and scalable. A natural question to ask next is whether we can design such compact routing schemes for the Internet.

The Internet is a complex network and very little may be said about its continuously evolving structure. It is widely accepted that the Internet Autonomous Systems (AS) level topology is *scale-free* i.e. the distribution $P(k)$ of node degrees k follows a power law $P(k) = k^{-\lambda}$ [56, 110], and it obeys the *small-world* property i.e. more than 80 % of the Autonomous System (AS) pairs are two to four hops apart [31, 95]. We shall elaborate more on small-world property when we introduce our second model in Sect. 5.3. The current Internet routing architecture suffers from several problems mainly inflexibility and scalability concerns. The former is due to the fact that addressing must be bound to topology and aggregation must be aggressive in order to achieve efficiency; both of these properties are broken by mobility, multi-homing, and in general de-aggregation of the address space [102] as discussed in the previous section. Scalability is questionable as well mainly due to the exponential growth in the Routing Information Base (RIB) which is growing much faster than the network itself is [88]. Some recent work has focused on remedying this problem [79, 152] but the work is limited in that scaling of the RIB is at best linear. Such work has considered routing on AS numbers (currently on the order of 10^5) rather than IP prefixes (currently on the order of 10^6). Ideally, a routing architecture should be able to provide topology-independent addressing and scaling at the same time, to which aggregation remains the only known effective solution and is unfortunately being broken by AS policies. Can we use the Abraham scheme for salable NICR on the Internet for example? The Abraham scheme performs poorly on the Internet due to its static nature; topology changes require updating the locator information to follow the topology in addition to updating the identifier-to-locator mappings. Indeed, separating the endpoint identifier from the routing identifier (locator) is a widely accepted evolutionary approach to scaling the Internet routing system and there has been significant work in that direction. Such Locator-Identifier Split (LIS) schemes for example (e.g. [66, 118, 120]) are a special case of name-independent routing which assumes aggressive aggregation to solve the routing table scalability problem among others. Similarly to the Abraham scheme, Internet LIS proposals are inherently less efficient than labeled compact routing schemes mainly due to the addition of identifier-to-locator mapping state and resolution overhead.

The question of whether routing architecture scalability is achievable on Internet-like graphs is investigated by Krioukov et al. in [102] in light of compact routing research. The authors show that hierarchical addressing is not efficient on the small-world Internet topology as the former requires large distances (n^{v}, $v > 0$) between nodes for efficiency. Two major points are emphasized for universal compact routing: (1) the first says that the lower space bound for shortest-path routing is $\Omega(n \log n)$ which is the same as the trivial case i.e. when each node keeps an entry in its Routing Table (RT) for each of the other $(n-1)$ nodes; and (2) the second surprising point is that there exists name-dependent and name-independent universal stretch-3 routing schemes with the same $O(\log n)$ RT size upper bound. Relaxing the universal routing restriction and focusing on the Internet scale-free graphs, the authors show how the Thorup–Zwick LCR scheme [157] is able to achieve an average stretch of 1.1 and an RT size of 50, thus significantly outperforming hierarchical schemes and routing on AS numbers schemes (e.g. [152]). Such observation has motivated some other work that provided infinite scaling schemes $O(\log^2 n)$ for static power-law graphs. The Brady–Cowen LCR scheme additionally is specialized for power-law graphs and scales infinitely [41]. As for dynamic graphs with topology updates, communication cost becomes an important factor to consider. Korman and Peleg [101] showed that the Internet scale-free graphs are the worst case graphs in terms of adaptation cost, scaling with $\Omega(n)$ messages, just the same rate as universal graphs. To summarize, the scale-free property of the Internet graph admits compact routing schemes that are highly efficient (low average stretch), but the same property results in worst case scaling in terms of adaptation cost. Designing scalable schemes when it comes to routing updates on scale-free graphs remains an open problem.

5.3 Navigating in a Small World

The compact routing model is fairly general as it has allowed us to derive useful tradeoffs across a wide range of networks. However, when working with complex networks, the model quickly becomes intractable if we are to derive meaningful results. Social networks, the world wide web, and the Internet Autonomous System (AS) level topology are examples of complex networks. These networks are extremely complicated that they just look random, and very little may be said about their structure beyond general laws [31]. In this section, we explore a more specific model for discovery in complex networks that exhibit the *small-world* phenomenon. Remarkably, discovery on small-world networks is feasible and even efficient using only local information. In social networks for example, people (nodes) need not keep a global map about the rest of the network; they only need to know their friends (neighbors) and some useful clues about their targets, such as occupation and/or geographic location, in order successfully and efficiently locate the targets. We start by motivating the small-world phenomenon, then we develop the model on social networks, and we explore some extensions of the model to Internet inter-domain routing.

The first empirical study to reveal the surprising small-world structure of social networks was conducted by the psychologist Stanley Milgram in 1967 [116]. Milgram showed how any two people in the world are likely separated by only a small number of intermediary connections, mainly 6 *degrees of separation*. In his experiment, Milgram gave a set of random people letters and asked them to forward these letters to their designated targets such that letters may not be sent directly to the target; each person has to advance the letter by sending it to a direct acquaintance, someone they know on a first-name basis. Each letter had additional clues about the target mainly the target's name, address, and occupational information. Remarkably, a third of the letters reached their final destination in a median of six hops. The astonishing result has triggered a lot of interest in further exploring the small-world phenomenon and several related experiments have been conducted since. The most recent and broad one studied the Facebook global social graph and showed strikingly that the world is getting even smaller [28]. The number of hops between all pairs of individuals on Facebook is found to be 4.78 instead of six degrees.[4] The experiments demonstrate how large social networks have abundant short paths. In addition, Milgram's experiment shows surprisingly that people can find these short paths without having a global view of the network (by only relying on local knowledge).

The efficiency of decentralized search (or discovery) in complex small-world social networks has been pioneered by Kleinberg [55, 94, 95]. Kleinberg studied the two parts of the problem: what causes the emergence of short paths, and under what circumstances can we efficiently find them? The abundance of short paths alone is not enough and not even necessary to admit efficient discovery. It is shown by Watts and Strogatz [162] that only a small number of random long-range weak links on a network could reproduce the small-world effect. What are the structural properties of these links then that allows people to efficiently find short paths? For example, when the long-range links are chosen at random, the discovery process is unable to find them. There is an important qualitative property of the links that results in efficient decentralized search. We state the property first and then explain it in more

[4]While 99.6 % of all pairs of users are connected by paths with five degrees (six hops), 92 % are connected by only four degrees (five hops).

detail. We start with a network with nodes clustered based on some underlying distance measure d whether geographic distance or social distance. This is natural to assume in social networks where people are linked to their geographic neighbors and friends that belong to the same social groups. If we then let each person/node choose a small number of long-range weak ties/links such that the probability of node u forming a link to node v is proportional to the distance $d(u, v)^{-2}$, then the small-world property emerges and, more importantly, discovery is shown to be efficient [94]. Using this inverse-square distance law forms links that are uniformly spread across different scales of resolution. To elaborate, if we draw a circle of radius d around node u, then the number of nodes that belong to this circular area is proportional to d^2, while the probability of forming a link to a node around the circumference of the circle is proportional to d^{-2}. This cancellation of exponents is what causes discovery to be independent of distance scale and hence be efficient. To recapitulate what we just explained, small-world complex social networks admit efficient discovery using only local information (clustering) and minimal knowledge about the network (in the form of weak long-range links) when certain qualitative properties hold. This result is remarkable as it means that we could potentially design highly scalable and efficient discovery schemes on such complex networks without having to construct a global map of the network. As usual, we now ask whether the results may be applied to the Internet specifically.

The recent work by Krioukov et al. on Hidden Metric Spaces [103] sets out to explore exactly that: can routing be performed efficiently on the Internet AS-level network topology without the need for exchanging global information about the connectivity graph? The authors show that under certain assumptions about the underlying topology, this is indeed feasible. A node needs to only know of (i) a global structure in which the network is embedded, and (ii) its locally connected neighbors. For example, a node knows its coordinates in some underlying metric space along with the coordinates of its direct neighbors. Routing then is simply forwarding to the neighbor that is closer to a destination in the metric space, a process referred to as *greedy routing*. Hence a "distance" metric is defined based on the structure (metric space) in which the network is embedded. Let us look at a simple example to better envision this process. Consider the simple network of Fig. 5.5 where nodes are placed on a grid and each node is connected to all its direct neighbors on the grid. Such a network may be directly embedded into a cartesian plane: each node is assigned a set of coordinates with the top left node say as the origin. Distance between two nodes with coordinates (x_1, y_1) and (x_2, y_2) is simply the Manhattan distance defined as $d((x_1, y_1), (x_2, y_2)) = |x_2 - x_1| + |y_2 - y_1|$. Given the coordinates (t_x, t_y) of a target node t, a node u simply forwards to a neighbor $v = (v_x, v_y)$ that is closer to (t_x, t_y) than itself in the Cartesian space i.e. such that $d((v_x, v_y), (t_x, t_y))$ is minimized over all neighbors v. In this example, each node only needs to know its coordinates and those of its neighbors (constant space) in order to find shortest paths. Obviously, the key challenge for the success of greedy routing lies in finding a "reflective" metric space on which greedy routing succeeds. The only property that is required of the metric space is the following: distance closeness in the metric space is equivalent to direct connectivity in the network topology with high probability. Some simple structures and relative metrics for efficient discovery, described

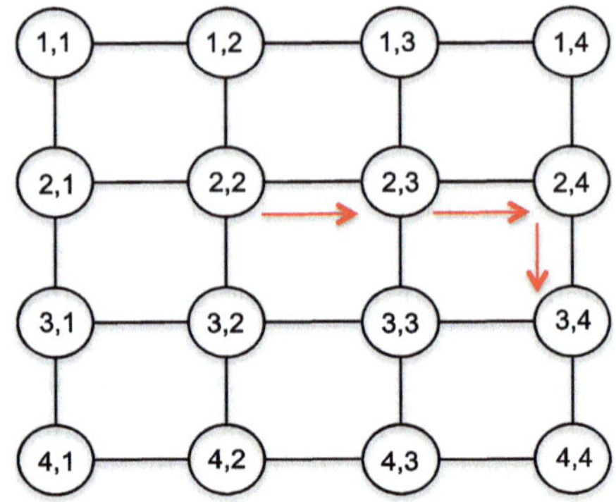

Fig. 5.5 Greedy routing on a grid network. The distance $d((2,2),(3,4)) = |2-3| + |2-4| = 3$. Node with coordinates $(2,2)$ forwards the message to a neighbor that is closer to $(3,4)$ than itself i.e. either $(2,3)$ as shown here or $(3,2)$, and the process repeats

earlier by Kleinberg [94], are the lattice, the hierarchy, and the set/group spaces. However, finding a metric space in which the Internet AS-level topology may be embedded seems like a much harder endeavor.

Boguna et al. present one such embedding of the Internet AS-level topology in a hyperbolic metric space [38], which build on an earlier work by R. Kleinberg [98]. The authors show how greedy routing is efficient on the complex networks that emerge from hyperbolic spaces even when the network is highly dynamic [126]. Nodes only maintain their coordinates in a hyperbolic metric space and forwarding in the space is performed using the distance gradient method described earlier. The node coordinates survive network dynamics mainly in terms of node and link failures. For details on the hyperbolic metric space and its congruency with the network topology we refer the reader to [38, 126, 142]. Greedy routing on hidden metric spaces may be thought of as a highly efficient, scalable, and dynamic labeled-compact routing scheme where node labels are chosen to be the coordinates in the metric space.[5] Clearly, these labels cannot survive mobility, and extending the scheme to a name-independent version necessarily requires an indirection layer on top of the greedy routing layer. In contrast, Milgram's experiment and Kleinberg's work on the navigability of social networks do not seem to have the same limitation: the names of targets are independent of the actual network structure and can accordingly be persistent and survive mobility. An interesting open problem is whether we can do efficient greedy routing on persistent names instead.

5.4 Conclusion

This chapter explored the design of name-based discovery systems in computer networks, mainly the Internet. We explored two models towards this end. The first is the compact routing model which is concerned with designing efficient routing schemes

[5] It is not clear, however, how such proposal might handle changes in the metric space itself or node dynamics beyond simple joins.

using compact distributed data structures. The model highlighted three properties of interest during the design of discovery schemes mainly efficiency, scalability, and adaptability, and explored the design space in terms of tradeoffs between stretch, space, and adaptation cost. The space vs. stretch tradeoff for example is directly associated with scalability vs. efficiency of the routing system. We shall leverage this observation in the next chapter to introduce the service differentiation concept i.e. routing schemes that provide different service levels to different sets of nodes. The compact routing model is general as it highlights the design tradeoffs for a wide range of networks. The main takeaway of the chapter is that the designer is working in a tradeoff space. We then showed how highly efficient and scalable discovery schemes may be designed if certain assumptions about the structure of the network hold. Motivated by efficient decentralized search in social networks, we reviewed one such model that leverages some "distance" property to design efficient greedy routing on complex large-scale networks.

Chapter 6
Discovery Service Differentiation

I don't know where I am going, but I am on my way
Voltaire

In Chap. 3 we defined names, and we differentiated locators as names that embed location semantics, whereas identifiers do not. We defined name-based discovery in Sect. 3.4 as the end-to-end process by which a message reaches its intended destination object starting with a name. When the starting name is a location-independent identifier, we refer to the process as identifier-based discovery instead. We explicitly distinguished between *discovery* and *routing* by stating that routing is the subprocess that acts directly on the main message while discovery additionally includes other mechanisms that assist in correctly forwarding the message without directly acting on it. We showed in Chaps. 2 and 4 examples of identifier-based discovery schemes that either utilize mapping/resolution where the identifier is mapped to some locator (e.g. [45, 66, 112, 118], and the Domain Name System (DNS) [117]), or utilize routing-on-identifiers (e.g. [19, 27, 42, 90, 99]). Chapter 5 then elaborated on the main tradeoffs involved in designing discovery schemes. Building on the discussion of Chap. 5, this chapter is an exercise in the design of *differentiated* discovery services. We first define and motivate a multi-level discovery problem that accounts for discovery service differentiation and we present an architecture that provides differentiated services.

6.1 Introduction and Motivation

A named object (such as a node or service), referred to as a player hereafter, demands to be discoverable by the rest of the network. A discovery scheme provides such service to the players. We define the *discovery level* to be a measure of "how discoverable" a player is by the rest of the network. This is "how easy" it is for the network to discover the player not the opposite. The performance of discovery, or the discovery level, could significantly affect the player's business model especially in time-sensitive application contexts. If discovering an entity takes a significant time

J.S. Khoury, C.T. Abdallah, *Internet Naming and Discovery*,
Signals and Communication Technology, DOI 10.1007/978-1-4471-4552-3_6,

relative to the delivery/download time, the experience of the requesting user suffers. As an example, when no caching is involved, the DNS resolution latency comprises a significant part of the total latency to download a webpage (10–30 %) [33, 87]. This overhead becomes more noticeable in Content Distribution Networks (CDNs), where content objects are extensively replicated throughout the network closer to the user and the discovery (or resolution) could potentially become the bottleneck. Traditionally, the design of discovery schemes has assumed that all players have the same discovery performance requirements, thus resulting in homogeneous demand. In such a setting, the discovery schemes deliver a discovery service that is oblivious to the actual, possibly heterogeneous, discovery requirements—and valuations—of the different players. In reality, however, the CNN site will likely value a higher discovery level more than a generic residential site. The main question posed in this chapter is therefore the following: should the design of discovery schemes account for service differentiation? We answer this question by introducing the *Multi-Level Discovery* (MLD) framework which is concerned with the design of discovery schemes that can provide different service levels to different sets of players. To further motivate the problem, we note that on the current Internet, Akamai provides such an expedited resolution service [14]. However, the service which is based on DNS suffers from the same pitfalls of the latter (expensive first lookup and critical dependence on caching) and tightly couples the content distribution provider with the resolution service provider.

The first question we ask is whether differentiated discovery is algorithmically feasible i.e. is it possible to devise a scheme that is scalable and that provides different levels of service to different players? Along this dimension, we define the algorithmic problem in Sect. 6.2, and we present a proof-of-concept MLD scheme in Sect. 6.3 along with an analysis of its scalability properties.

6.2 What is Multi-Level Discovery (MLD)?

We start by providing a generic definition of the MLD problem. The problem specifics will depend on the context, mainly the design assumptions and requirements.

Definition 6.1 Multi-level discovery (MLD) problem statement: Given a graph $G = (V, E)$, a set of nodes with unique identifiers (identifier of node i is simply $\langle i \rangle$), set of m discovery levels where each node is associated with some level $l \in \Lambda$,[1] devise a discovery scheme that routes on identifiers. The set Λ of possible discovery levels is known to all nodes. The scheme is expected to deliver to each node i in G its requested discovery level $l \in \Lambda$.

The main challenges inherent to the MLD problem arise from the following requirements:

[1] When the set of discovery levels is discrete, a level becomes a "class" of service.

- different levels of service must be supported by the same scheme, and
- the discovery level of a destination $\langle t \rangle$ is unknown at the time of discovery. The challenge here is that information about the discovery level of the destination is to be discovered as well by the scheme and is not known a priori. The only attribute that is known a priori is the identifier.

6.3 A Multi-Level Discovery Scheme

A traditional class of discovery schemes that has been extensively investigated in the research community is the general Name Independent Compact Routing (NICR) problem first introduced in [27]. We have introduced the NICR problem previously in Chap. 5 and we have reviewed Abraham's optimal NICR scheme on universal graphs [19]. NICR is of particular interest to this section and we shall extend the framework for implementing a MLD scheme. We restrict our attention to trees rather than universal graphs. More specifically, we extend Laing's NICR scheme [106] which operates on top of the optimal Thorup–Zwick labeled routing scheme on trees [157]. The latter represents a locator-based routing function over which the identifier-based discovery scheme is implemented.

6.3.1 Background: NICR Scheme on Trees

A name-independent compact routing scheme on trees (NICRT) is developed by Laing [106] with a space/stretch tradeoff based on a parameter k. The scheme achieves stretch $2^k - 1$ for a space requirement of $\tilde{O}(k^2 n^{1/k})$, where n is the number of nodes. From a high level perspective, the tradeoff is achieved by asking each node to know about a set Σ^i of nodes ($|\Sigma^i| = n^{i/k}$) at concentric circles or neighborhoods $N^i, 0 \leq i \leq k-1$ from itself. Routing towards a destination d proceeds through prefix matching of d's identifier $\langle d \rangle$ represented in base $n^{1/k}$ (denoted by $\langle d \rangle_{n^{1/k}}$). Delivery is guaranteed in at most k hops i.e. by matching the k letters of d's identifier base $n^{1/k}$. The main idea is that as the value k increases (i.e. as the number of concentric circles or layers increases), a node will keep less information about the rest of the network but the stretch which is directly proportional to the number of layers will increase. On the other hand, as k decreases (i.e. fewer layers), a node will keep more information about the rest of the network and the stretch decreases accordingly.

Laing's scheme is based on a coloring theorem for trees. The coloring theorem states that any tree with n nodes can be colored with q colors such that every neighborhood $N_q(v)$ of size q (for every node $v \in V$) is distinctly colored i.e. each node in $N_q(v)$ has a unique distinct color from the set of colors $[q]$ (check [106]). For reference, we include the theorem here:

Theorem 6.1 [106] *Let $T = (V, E)$ be a tree with $n \geq q$ nodes ($q \geq 1$), and positive edge weights. There exists a function $c : V \rightarrow [q]$ such that $\forall v \in V$, $N_q(v)$ is fully colored.*

The theorem is used in the NICRT scheme to uniquely color neighborhoods $N^i(v)$ of size $n^{i/k}$ at each layer $i, 0 \ldots k-1$.

Laing's scheme works as follows: Given a tree $T = (V, E, w)$, and a $k \geq 1$, multiple layers of coloring are assigned to nodes as follows: at layer $1 \leq i \leq k-1$, T is fully colored with Σ^i colors where $|\Sigma^i| = n^{i/k}$ and $\Sigma = \{0, 1, \ldots, n^{1/k} - 1\}$ is the alphabet. Note that the neighborhood of a node v is denoted by $N^i(v)$ and is the set of $n^{i/k}$ closest nodes to v including the latter. Hence $|\Sigma^i| = |N^i|$ and the coloring theorem achieves a full coloring. Each node $u \in V$ is hence assigned a unique color $c_i(u)$ at layer i, where $c_i(u) \in \Sigma^i$. In addition to the $k-1$ colors node u obtains, it has its unique identifier $\langle u \rangle$ picked from the set $\{0, \ldots, n-1\}$ and represented in base $n^{1/k}$ and padded to the left with zeroes. Thus $|\langle u \rangle| = k$.

Storage Each node u has an identifier $\langle u \rangle$ and $k-1$ colors $c_i(u)$. Denote by $\sigma_i(u)$ the length i prefix of $\langle u \rangle$. In addition to the labeled compact routing table information of [157],[2] node u creates its routing table according to Algorithm 6.1.

Routing In terms of routing to some destination t with identifier $\langle t \rangle$ starting at some source s, routing proceeds as indicated in Algorithm 6.2.

Note in Algorithm 6.2 that each next hop (i.e. v_{i+1}) is guaranteed to belong to $N^{i+1}(v_i)$. Note as well that the only node that matches $\sigma_k(t)$ is the node whose identifier is $\langle t \rangle$ which guarantees delivery [106].

Algorithm 6.1 Routing table construction for node u

for each layer $i, 0 \ldots k-1$ **do**
  Let $\lambda = \{c_i(u), \sigma_i(u)\}$, where $c_0(u)$ and $\sigma_0(u)$ are the empty string ϵ
  for each $\tau \in \Sigma$ **do**
    store label of closest node v to u that satisfies $c_{i+1}(v) = \lambda\tau$ or $\sigma_{i+1}(v) = \lambda\tau$
  end for
end for

Algorithm 6.2 Routing to $\langle t \rangle$

let $v_0 = s$
for each layer $i, 0 \ldots k-1$ **do**
  route to node v_{i+1} which is the closest node to v_i that matches $\sigma_{i+1}(t)$ i.e. node v_{i+1} satisfies $c_{i+1}(v_{i+1}) = \sigma_{i+1}(t)$ or $\sigma_{i+1}(v_{i+1}) = \sigma_{i+1}(t)$
end for

[2] This information is used for optimal stretch-1 routing based on locators (topological labels).

6.3.2 Extending Laing's Scheme

In the preceding scheme, the effect of the parameter k was to control the space/stretch tradeoff achieving stretch $2^k - 1$ for a space requirement of $\tilde{O}(k^2 n^{1/k})$. In this section, we extend Laing's scheme by allowing multiple stretch levels (or multiple values of k) on the same tree T for different sets of nodes. Discovery levels will correspond to values of k in Laing's scheme which directly determines the stretch.

More clearly, we assume the existence of a set $K = \{k_1, \ldots, k_m\}$ $(m = |\Lambda|)$ of stretch levels ordered in ascending order with $\Lambda \subseteq \mathbb{Z}^+$. Assume also without loss of generality that n is a k_mth power and that $k_1 \geq 2$. Each k_i corresponds to a discovery level $l = \frac{1}{2^{k_i}-1}, l \in [0, 1]$ and we assume that $m = |\Lambda| = O(n^{\frac{1}{k_m}})$. The main idea that we shall use for extending Laing's algorithm to support multiple discovery levels on the same tree T introduces ACCELERATE tables that expedite discovery/routing for nodes that demand higher discovery levels. The extended scheme starts by providing the lowest discovery level $(\frac{1}{2^{k_m}-1})$ to all nodes by constructing Laing scheme for $k = k_m$. The pseudocode for construction of the routing tables is listed in Algorithm 6.3. Lines 6, 17 in Algorithm 6.3 and lines 5, 6 in Algorithm 6.4 encapsulate the main logic for expedited discovery.

Algorithm 6.3 Extended table construction for node u

1: Let $K' = \{k_1 - 2, \ldots, k_{m-1} - 2\}$
2: **for** each layer $i, 0 \ldots k_m - 1$ **do**
3:     Let $\lambda = \{c_i(u), \sigma_i(u)\}$, where $c_0(u)$ and $\sigma_0(u)$ are the empty string ϵ
4:     **for** each $\tau \in \Sigma$ **do**
5:         **if** $i \in K'$ **then**
6:             store label of closest node v to u that satisfies $c_{i+1}(v) = \lambda\tau$
7:             store label of closest node w to u that satisfies $\sigma_{i+1}(w) = \lambda\tau$ only if $w \in N^{i+1}(u)$
8:         **else**
9:             store label of closest node v to u that satisfies $c_{i+1}(v) = \lambda\tau$ or $\sigma_{i+1}(v) = \lambda\tau$
10:         **end if**
11:     **end for**
12: **end for**
    {Construct the ACCELERATE table}
13: **for** each level $k_j, j : m - 1$ down to 1 **do**
14:     Let $s = k_j - 1$
15:     Let D_{k_j} be set of nodes requiring level k_j
16:     **for** each node $u \in D_{k_j}$ **do**
17:         Let B_s be set of nodes whose color at layer s is $\sigma_s(u)$
18:         Add extra pointer $\{\langle u \rangle_{n^{1/k_m}} \rightarrow \mathit{label}(u)\}$ at each node $v \in B_s$
19:     **end for**
20: **end for**

Algorithm 6.4 Routing to $\langle t\rangle$ using extended scheme

let $v_0 = s$
for each layer $i, 0 \ldots k_m - 1$ **do**
  if $\langle t\rangle \in$ ACCELERATE table **then**
    route directly to t using *label*(t)
  else if $i = k_j - 2$ for any $j = 1 \ldots m$ **then**
    route to node v_{i+1} which is the closest node to v_i such that $c_{i+1}(v_{i+1}) = \sigma_{i+1}(t)$ or $\langle v_{i+1}\rangle = \langle t\rangle$
  else
    route to node v_{i+1} which is the closest node to v_i that matches $\sigma_{i+1}(t)$ i.e. node v_{i+1} satisfies $c_{i+1}(v_{i+1}) = \sigma_{i+1}(t)$ or $\sigma_{i+1}(v_{i+1}) = \sigma_{i+1}(t)$
  end if
end for

In terms of routing to destination $\langle t\rangle$ using the extended scheme, we extend routing Algorithm 6.2 as depicted in Algorithm 6.4 given that each node knows the set of stretch levels k_j, $j = 1, \ldots, m$.

Analysis It can be easily verified that delivery is guaranteed as well as $d(v_i, v_{i+1}) \le 2^i d(s,t)$ in the extended algorithms (check [105]). In order to maintain the sub-linear space requirements at each node, the extra state maintained at each node for discovering higher level nodes must be less than a constant factor of $k^2 n^{\frac{1}{k}}$. First, at line 15 of Algorithm 6.3, in the worst case there are at most $n^{1-\frac{s}{k_m}}$ nodes in D_{k_j} that have the same length s prefix (when $|D_{k_j}| = n$) i.e. that can potentially introduce state on the same set of nodes B_s. Thus the maximum increase in any node's routing table size is $m \cdot n^{1-\frac{s}{k_m}}$. We have already assumed that the total number of levels $m = O(n^{\frac{1}{k_m}})$. Formally, in order to maintain sub-linear space at each node, the following condition must hold: $n^{1-\frac{s}{k_m}} \le \alpha k_m^2 n^{\frac{1}{k_m}}$ for some large constant α, or $s \ge k_m(1 - \frac{\log \alpha k_m^2}{\log n}) - 1$. This constraint must hold when choosing the set of possible discovery levels Λ (and hence the respective set K) in order for the extended routing scheme to satisfy the sub-linear space requirement inherent to compact routing design.

6.4 Discussion and Conclusion

The MLD framework allows for discovery service differentiation. We have defined the problem, motivated it, and demonstrated its algorithmic feasibility in the context of NICR. Our interest in compact routing in this chapter is primarily due to the mathematical bounds imposed on space and stretch which support our discovery level concept by providing guarantees on levels of performance. While we have studied the problem for NICR, differentiation of discovery is important as well in the context of the TCP/IP Internet. All current discovery schemes (such as DNS) suffer from the same problem: performance of discovery. If we ever think of using domain names as endpoint identifiers in the TCP/IP Internet, the impact of discovery (i.e. the first mapping from domain name to IP address) becomes of great importance. This is due to the fact that TCP treats the first packet as representative of congestion.

Finally, in this chapter we have focused on the algorithmic feasibility questions. Notice, however, that there is a non-trivial cost associated with being discoverable. This could be the cost of distributing and maintaining information (state) about the identifiers to provide a certain discovery level. Hence, the second challenge is that of providing an economic model that accounts for cost and valuation in the design of discovery mechanisms. The next two chapters (Chaps. 7 and 8) are dedicated to studying the economic dimension. By adding an economic dimension to the discovery design space, we hope to gain more knowledge about the complex design decisions pertaining to naming and discovery in networks, and to be able to design discovery mechanisms that are suitable for a future Internet.

Part III
Economics of Discovery

Chapter 7
On the Economics of Identifier-Based Discovery

If you have to forecast, forecast often
Edgar R. Fiedler

Chapter 5 introduced the discovery function and discussed the design tradeoffs as they relate to efficiency, scalability, and adaptability. In this chapter and the next we study the economic dimension. This chapter presents a broad treatment of the main economic issues that arise in the context of name-based discovery. Providing a discovery service while accounting for the cost and making sure that the incentives of the players are aligned is the general theme. We motivate the subject, and we present a taxonomy of discovery schemes and proposals based on their business model.

7.1 Introduction

Traditionally, the design process in the context of the Internet has focused on sources of value as they relate to performance, robustness, resilience, reliability, etc. with less emphasis on the socio-economical dynamics. The value of any new design in the new era does not solely depend on performance and must take into account the complex social and economic interactions and incentives of the agents using the design if success is to be reached [96, 125]. Check [125] for an interesting overview of several tools that are important in bridging computer science and economics to better understand the complex socio-economic interactions in the context of the Internet, and [96] for an interesting overview of several of the problems and applications arising at the interface between information and networks.

The previous chapters have motivated the importance of naming and discovery in computer networks. We assume a naming system for the Internet is required given the network's mobile and ubiquitous usage models. For example, on the Internet, this translates into either designing a new system or enhancing the current ones (for example DNS). While there is a rich literature on applying game theory and economics models to Internet games, we find in the networking literature a number of

J.S. Khoury, C.T. Abdallah, *Internet Naming and Discovery*,
Signals and Communication Technology, DOI 10.1007/978-1-4471-4552-3_7,

proposals for Internet discovery schemes requiring significant coordination among selfish users while ignoring the economic aspects that may possibly render them infeasible or inefficient (and we shall give several examples of such systems or proposals later in Sect. 7.3). In a future Internet in which domains or Autonomous Systems (ASs) are self-interested, utility-maximizing agents, the design of any identifier-based discovery scheme could benefit from establishing the right economic models. The problem on the Internet specifically is exacerbated as there are multiple layers of identification managed by different systems, such as DNS [117] at the application and the Border Gateway Protocol (BGP) [132] at the network layer.

In Chap. 5, we have introduced discovery in large-scale networks. We have additionally defined the *multi-level discovery* framework in Chap. 6 which is concerned with the design of discovery schemes that can provide different service levels to different sets of nodes. Obviously, there is a cost associated with being discoverable. This could be the cost of distributing and maintaining information (state) about the identifiers. In current schemes, the discovery demand is actually insensitive to cost since no cost structure exists and hence demand flattens out to a homogeneous level. Accounting for and sharing the cost of discovery is an interesting problem whose absence in current path discovery schemes has led to critical economic and scalability concerns. As an example, the Internet's BGP [132] control plane functionality is oblivious to cost. A BGP speaker that advertises a provider-independent prefix (identifier) does not pay for the cost of being discoverable. Notice here that BGP with its control and forwarding planes represents a discovery scheme on prefixes which are technically flat identifiers in a largely de-aggregated namespace. Hereafter, we refer to this form of BGP as BGP-DA for De-Aggregation. This problem becomes more important in settings where the state (and the cost) is incurred by service nodes that are not themselves players.[1] In this case, the cost must be paid for or else the service nodes will have no incentive to implement the discovery service. Hence, *we conjecture that a discovery scheme should be aware of incentives and cost necessitating that players/nodes pay for the cost of getting the service*. Providing such a service while accounting for the cost and making sure that the incentives of the players are aligned is the general economic problem that we frame in this chapter.

The rest of the chapter is organized as follows: first we review background material in Sect. 7.2. Specifically, we motivate the notion of strategic interactions on networks by presenting three games in Sect. 7.2.1 that we shall refer to throughout the discussion. We also distinguish between the search function and receiver-based discovery function in Sect. 7.2.2. Distinguishing the two functions is important to frame our work. Section 7.3 presents a taxonomy of discovery schemes based on their business models. Finally, Sect. 7.4 presents our thoughts on suitable economic models for the different discovery models highlighted by the taxonomy before concluding in Sect. 7.5.

[1] Service nodes implement the discovery service. Players are customers of the discovery service or agents that wish to be discoverable.

7.2 Background

7.2.1 Networks and Strategic Behavior

Game theory is a fundamental mathematical tool for understanding the strategic interactions among selfish network agents, particularly on the Internet over which self-interested agents (e.g., ASs) interact. The theory provides several solution concepts to help study games that arise in different situations and that have specific requirements and varying underlying assumptions [123]. We overview some basic ones here and we provide examples to illustrate each. The most central and widely applicable solution concept is the *pure strategy* Nash equilibrium (PSNE or NE) which could be simply thought of as a set of strategies that forms a stable solution of the game. A set of strategies for the players is termed a *strategy profile*. Under NE strategy profile, no player can deviate profitably (i.e., increase her payoff) from her strategy given every other player's strategy. Despite its wide applicability, the NE solution has several shortcomings in that it may not exist (and hence might require mixing), there could be multiple equilibria, and it might be computationally intractable to get to it. In this sense, the *mixed strategy* solution concept was developed by Nash. Mixing the strategies guarantees that an equilibrium always exists in the game by introducing probability distributions over the pure strategies (hence rendering the strategy space a convex set). A more stringent solution concept is the *dominant strategy* solution. Unlike the pure strategy solution, a dominant strategy yields a player the highest payoff independent of the strategies of the rest of the players. Dominant strategies are very attractive solutions when they exist, and when they do not exist, game designers might try to design for them. For example, when a player's strategy is to declare some private information that is necessary to the social welfare of the game, an attractive solution would be to make the truthful revelation a dominant strategy. This makes sure that the player does not have an incentive to lie. The mechanism design framework [111] provides exactly this solution allowing the mechanism "designer" to achieve a dominant strategy solution (in addition to other design goals). An extension to mechanism design, Algorithmic Mechanism Design (AMD) [122], deals with the computational complexity of the solution and Distributed AMD [60] further considers the "network complexity" in distributed settings. Several other solution concepts exist; however, we only review one more which is the *subgame perfect* Nash equilibrium (SPNE). SPNE extends the one-shot NE concept to settings in which players take turns playing (e.g., player 1 plays first, then player 2 plays). In such a setting, SPNE becomes more natural as it captures the order of decision taking. Briefly, a SPNE is a NE in every subgame of the original game, where a subgame may be informally defined as a portion of the game that can be independently analyzed. Note that by the formal definition of a subgame, every game is a subgame of itself and hence every SPNE is necessarily a NE. For formal definitions of the solution concepts and a comprehensive treatment of the topic, we refer the reader to [65].

How does strategy factor into networking problems? To motivate the importance of strategic behavior, we hereby present three networking applications that employ different solution concepts and that we shall refer to throughout the discussion. Our hope is that the games highlight some of the basic economic issues that are of interest to network settings and the tools that are directly applicable to these settings. Note that the games we present here might not be straightforward for the unexperienced reader who we refer to [65, 123] for introductory material on the subject. The first application we present is that of "query incentive networks" and is due to Kleinberg and Raghavan [97]. The second application is that of "trading networks with price setting agents" due to Blume et al. [36]. The common aspect of the first two games is that price setting is a strategic behavior of the players which is not the case with the third application we present, "Incentive-compatible interdomain routing" due to Feigenbaum et al. [59]. Additionally, while the first two games are solely interested in studying the equilibria, the third presents a distributed mechanism that achieves the solution.

Nash Equilibria and Query Incentive Networks Game [97] Query incentives are motivated in peer-to-peer and in social networks where some root node, the seeker, issues a query seeking a piece of information on the network. The seeker does not know which nodes on the network have the answer neither does any other node. The only way to find the answer is to propagate the query deeper into the network until a node with an answer is reached. In order to do so, every node needs to incentivize its direct children to propagate the query deeper. Hopefully, a destination node with an answer is eventually reached. Propagation is assumed to occur on a tree and incentives are provided by each parent in the tree to its children in the form of rewards. A node that gets offered a reward will itself offer a smaller reward to its children if its does not posses the answer hence pocketing some reward if an answer to the query is found under the node's subtree. We shall refer to this game hereafter as the QUERY-GAME, and we note that this game is based on a similar game initially introduced by Li et al. [109].

Formally, each node (player) u receives a reward r from its parent and offers a reward $f_u(r) < r$ to its children if it does not have the answer. Otherwise, if u has the answer to the query it responds to its parent with the answer. Each node holds the answer with probability $1 - p$ and on average one in every n nodes holds the answer (n is referred to as the rarity of the answer). The node's strategy is hence $f_u(r)$ which is assumed to be integer-valued and the payoff is simply $(r - f_u(r))\alpha_u(\mathbf{f})$ where $\alpha_u(\mathbf{f})$ is the probability that an answer is found in the subtree rooted at u given that node u has played f_u and every other node's strategy is given by $\mathbf{f} = \{f_v, \forall v\}$ ($\mathbf{f}$ is a *strategy profile*). Figure 7.1 depicts a sample game on a tree.

There are several questions that arise in such a game: How will a node act strategically to tradeoff its payoff and the probability that an answer is found in its subtree? A higher promised reward potentially means higher probability of finding an answer but less payoff. How much initial investment r^* is required as a function of the tree structure and the rarity of the answer, denoted by n, in order to find an answer with high probability? The authors answer these questions in [97] by modeling a general class of branching processes parametrized on the branching factor b. The latter is the mean number of active offsprings (or children) per node in the tree

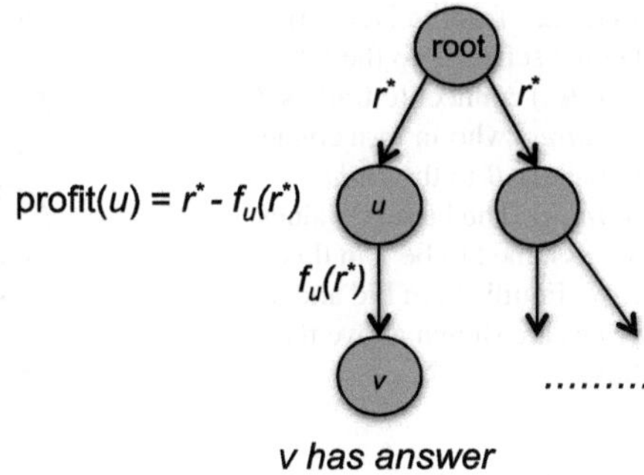

Fig. 7.1 Query Incentive Game: root offers a reward r^*, node v has an answer to the query

constructed using a random branching process [97] (when $b < 1$, the tree is almost surely finite while it is infinite when $b > 1$ with positive probability). When looking for the equilibria, one important point to notice in this game is the interdependency of the players' strategies as given by the tree structure; the strategy of a player depends on the strategies of its children and so on. The authors show that the Nash equilibrium exists (and is unique with some caveats) by constructing a set of functions $\mathbf{g}$ (a strategy profile) inductively and showing that the resulting strategy profile is indeed an equilibrium. This result simply says that there exists a stable solution $\mathbf{g}$ to the game such that if the nodes play the strategies $\mathbf{g}$ then no node is able to deviate profitably given the strategy profile of the rest of the nodes. However, the model does not provide a recipe to get to the solution. Knowing that a solution exists, the next step is to study the breakpoint structure of rewards to be able to say something about the initial investment required (check [97] for results there). In summary, the goal of this game (and the one in [109]) is to provide incentives for query propagation in decentralized networks with uncertainty about the destination of the answer. Such a process incurs cost that must be paid for by someone to keep the incentives aligned. The next game we discuss is based on the SPNE solution concept.

Subgame Perfect Nash Equilibria and Trading Networks Game [36] Trading networks, despite being more motivated from a markets angle, provide several insights into networking games that involve competition. A set of sellers S wish to sell their goods to a set of buyers B indirectly through a set of traders T. While [36] studies both cases where the goods are distinguishable or not, in this brief overview we shall only focus on indistinguishable goods, i.e., a single type of good where all copies are identical. Each seller holds exactly one copy of the good initially and each seller is only interested in buying one copy of the good as well. Trade between the buyers and the sellers may only happen through a set of traders T as specified by a graph G. G specifies how sellers and buyers are connected to the traders where each edge in G connects a node in $B \cup S$ to a node in T. Sellers are assumed to have zero value for the good while each buyer j has a value θ_j for the good. Figure 7.2 depicts such a setting where the indices i, j, t are used to refer to the sellers S, the buyers B, and the traders T, respectively.

We shall refer to this game as the TRADE-GAME. The game aims at studying the process of strategic price setting in markets with intermediaries, and proceeds

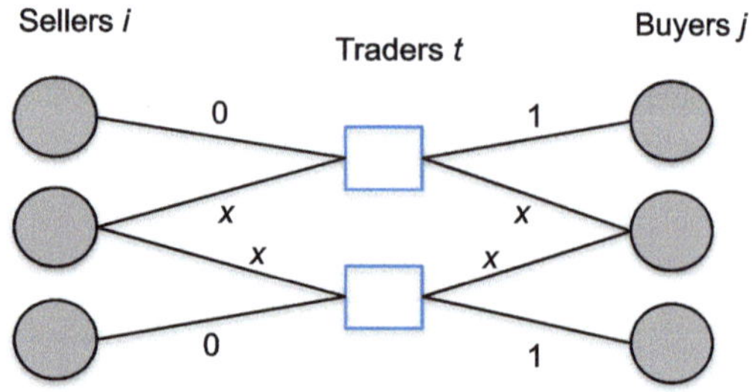

Fig. 7.2 Trading Network Game: sellers S to the *left* (*circles*) connect to traders T (*squares*) who in turn connect to buyers B to the *right* (*circles*). The buyers' values are assumed to be 1 in this case. Equilibrium bid and ask prices are shown above the links

as follows: first each trader offers a bid price β_{ti} to each seller i to which it is connected, and an ask price α_{tj} to each buyer j to which it is connected. The vector of bid/ask prices is the strategy profile of the traders. Then buyers and sellers choose among the offers they got, the traders pay the sellers the bid price and get the ask price from the buyers. If a trader gets more buyer offers than the seller offers it has, the trader will have to pay a very large penalty. This is so that such a scenario will never happen at equilibrium. The payoffs of the different players are as follows: a player that does not participate in a trade gets no payoff. A buyer that participates in a trade through some trader t gets a payoff of $\theta_j - \alpha_{tj}$, while a seller i that participates in a trade with trader t gets a payoff of β_{ti} (again here assuming the seller has no value for keeping the good). Finally, a trader that participates in trade with a set of buyers and sellers gets a payoff of $\sum_r (\alpha_{tj_r} - \beta_{ti_r})$ minus a penalty if more buyers than sellers accept its offer (where the index r runs for each distinct buyer, seller combination that have accepted t's offer). It is important to notice that price setting in this game is strategic. Hence, as in the previous game, the first question to ask is how will the traders act strategically to set the market prices knowing that multiple traders could be competing for the same business, and what solution concept is most suitable to studying this game? The solution concept used in this game is the subgame perfect NE which is suitable in such a two stage game where traders play first and then buyers and sellers react. With this in mind, the next step to understanding the strategic behavior of the players (or equivalently the price setting dynamics) is to ask whether a solution (equilibrium) exists and to understand the structure of any such solution. In Fig. 7.2, the equilibrium strategies are shown above the links. Two interesting equilibrium phenomena arise in this game: the effects of monopoly, and perfect competition. Both traders in this example make a maximum profit (of 1) from the single monopolized buyer/seller pairs that have access to one trader, while the traders make zero profit when competing for the business of the middle seller and buyer. This must be the case at equilibrium. It turns out as shown by the authors that the equilibrium always exists and that every equilibrium is welfare maximizing (where the welfare of an outcome is simply the difference between the values of the buyers and those of the sellers). These results are shown by resorting to the primal/dual solutions of a welfare maximization linear program. In any solution, no trader is able to make any profit unless the latter is essential to the social welfare of the game (this result captures the case where traders could have different costs and hence only the cheaper ones are part of the equilibrium). The game (with distinguishable goods) could be directly extended to account for trading costs, where traders incur costs to perform the trade. The same results hold whereby a trader is able to make profit only when the trader is crucial to the social welfare.

Mechanism Design and Interdomain Routing Game [59] The third game we present in this section is that of interdomain routing incentives, particularly for BGP. First, we briefly overview how BGP operates after which we proceed to describe the incentive mechanism. The Internet is mainly composed of independent Autonomous Systems (ASs), or administrative domains. ASs must coordinate to implement a distributed routing algorithm that allows packets to be routed between the domains to reach their intended destinations. BGP is a policy-based path-vector protocol and is the de-facto protocol for Internet interdomain routing. The protocol's specification [132] was initially intended to empower domains with control over *route selection* (which path or route to pick among multiple advertised routes to a destination), and *route propagation* (who to export the route to among the direct neighbors of an AS). The commercialization of the Internet quickly transformed ASs into economic entities that act selfishly when implementing their internal policies and particularly the decisions that relate to route selection and propagation [43]. Intuitively, selfishness and the lack of coordination could potentially lead to instabilities in the outcome of the protocol, as is actually the case with BGP. Griffin et al. have studied this problem and the authors provided the most widely accepted formulation, the stable paths problem. They provide sufficient conditions under which the protocol converges to a stable solution, the *no dispute wheel* condition [76]. In addition to the algorithmic side of BGP which deals with convergence and stability, recent work has focused on the economic side, particularly studying the equilibria of a BGP game and trying to align the incentives of the players (check [59, 108] and references therein).

The interdomain routing incentive game of [59], hereby referred to as ROUTING-GAME, aims to study the policies (strategies) under which BGP is welfare maximizing (i.e., it maximizes the social welfare), and incentive-compatible (i.e., no player has an incentive to deviate from telling the truth where the player's action is to declare private information). The goal is to design a distributed mechanism that provides these attractive properties. Formally, in this game we are given a graph $G = (N, L)$ that represents the AS level topology (nodes N are the ASs and L the set of links between them). The route computation problem is studied for a single destination d and may be directly extended to all destinations assuming route computation is performed independently per destination. Hence, there exists a set of n players indexed by i, and the destination d. Each player has a valuation function $v_i : P^i \to \mathbb{R}$ which assigns a real number to every permitted route to d, P^i being the set of all permitted routes from i to d. Note that a route is permitted if it is not dropped by i and its neighbors. No two paths are assumed to have the same valuation. Social welfare of a particular outcome, an allocation of routes R_i, $\forall i$ that forms a tree T_d, is defined to be $W_{T_d} = \sum_{i=1}^{n} v_i(R_i)$. Clearly, the concept of internal policy is captured with the strict valuation or preference function v_i over the different routes to d which is private information given that the nodes are autonomous. The goal is to design a mechanism that maximizes the social welfare despite the fact that its components, the v_i functions, are unknown or private. The mechanism design framework and particularly the Vickery–Clark–Groves (VCG) mechanism provides the solution [123]. To do so, a central bank is assumed to exist whose sole task is to allocate a payment $p_i(T_d)$ to each node i based on the outcome tree T_d. More clearly, a player may either truthfully reveal her valuation to the mechanism (by always picking the best valued routes to d) or not, hoping to manipulate the outcome to her advantage. Based on the players' actions and hence on the outcome tree T_d, a payment $p_i(T_d)$ is made by the central bank to each player. The utility of each player from an outcome is $u_i(T_d) = v_i(R_i) + p_i(T_d)$. The VCG payment scheme is intentionally designed to make the truthful action a *dominant strategy* for all players, hence no player has an incentive to lie about her valuation. To achieve this, AS i s compensated an amount p_i proportional to the decrease in the value of all upstream ASs that have picked their best route to d through i when the latter does not participate. This is exactly the impact on the social welfare when i is not playing [123]. From a game standpoint, the solution concept that was targeted is the dominant strategy solution—playing truthfully is a dominant strategy and achieving such an attractive solution comes at the expense of assuming a central bank that regulates payments. The authors show that BGP augmented with a VCG payment scheme is incentive-compatible and welfare maximizing in several well studied settings (assumptions on policies or valuation functions).

In the above problem, and generally in problems involving mechanism design, the common scenario is an allocation mechanism that distributes some resource to a set of participating players. In order for a mechanism to implement the Social Choice Function (SCF), for example maximizing the social welfare of all players, the mechanism needs to know the real private information (such as true valuations for example) of the players. This is the case because players might be able to strategically manipulate the output of the mechanism by lying about their private information or strategies. Hence, the mechanism tries to make "truthfully declaring the private information" a *dominant strategy* for the players.

7.2.2 Discovery versus Search

Before discussing the economic issues that arise in discovery mechanisms, we review two main operations in discovery and we discuss the source of value of each operation. We introduce the notions of *advertisers* and *seekers*. In identifier-based discovery, advertisers are the entities that wish to be discoverable by the rest of the network using their identifiers. They utilize the *join*(i, level) interface to express their wish to the mechanism. Seekers wish to locate the advertisers and they utilize the *discover*(i, j) operation to do so. In our model, players are advertisers who may simultaneously be seekers.

It is important to distinguish two different classes of problems that relate to discovery and that have been considered in the literature. The first, distributed information retrieval, is that of locating information without prior knowledge of the providers or the location of the information (information could be located anywhere in the network). This problem is generally referred to as unstructured search (as in Gnutella, Freenet P2P networks, social networks, etc.). One key idea here is that in order for the seeker to find the requested information, she must search for it and be willing to invest in the search. The provider either cannot or is not willing to do so. Prominent work in this vein that addresses cost and incentive structures includes the work by Kleinberg [97].

The second class of problems, which we are more interested in and which we refer to as identifier-based discovery, aims at discovering a path to a uniquely identified entity assuming the seeker is given the identifier(s) of the destination beforehand. This problem is common in service-centric networks where there generally exist many competing providers for the same service. Within this class of problems, we distinguish two subclasses based on the cost model employed. The first subclass deals with routing problems and focuses on the transit or forwarding cost which is to be bore by the seeker. Several proposals fall under this subclass and many utilize economic tools based in mechanism design [57, 122, 167]. We distinguish another flavor of the problem by noticing that in service-centric network environments, the seeker gets no utility from the discovery part but rather gets the utility from consuming the service itself. In this sense, the utility of discovery is mainly to the provider or the advertiser: the provider wishes to sell the service and can efficiently do so only when the service is "discoverable". This is the main point that distinguishes the models of this chapter from the literature on routing and forwarding incentives. The players may be thought of as providers that receive a utility from being discoverable by the rest of the network, the utility of being famous, the latter being inevitably related to the player's business. Hence, in the receiver-based business model, the player does not care about whether other players are discoverable or not, whereas with general P2P resource sharing applications the player's utility is to share the resources of other players and hence to be able to discover the rest of the network (originator-based).

7.3 A Taxonomy of Discovery Schemes

Figure 7.3 shows some classic models used by current discovery schemes (and proposals). Big circles (light and dark) represent nodes used by the routing function (nodes V). Big dark circles represent a subset of those nodes that maintains state about the identifier space. We refer to these nodes as the service nodes denoted with V_D where $V_D \subseteq V$. Small dark circles (colored red) are the objects, alternatively the players P, that wish to be discovered. Figure 7.3 tries to illuminate the relationship between the players P (who receive the discovery service), and the nodes V_D (who provide the service and incur the cost). This relationship is important in an economic setting, such as when studying pricing schemes and when devising a strategic model (and solution concept) for the problem at hand. For example, service nodes in model I (described shortly) may be generally considered to be obedient

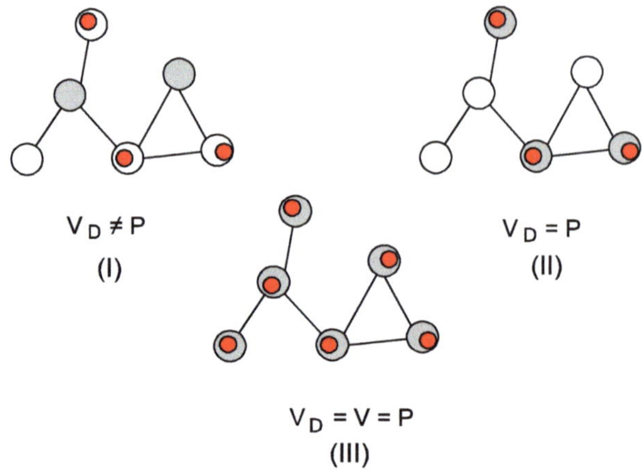

Fig. 7.3 Representation of some common models for discovery

Table 7.1 Identifier-based discovery schemes

Model	Representative schemes
model I	DNS, DONA [99], eFIT [112], LIS ([66], etc.)
model II	DHTs (Chord [150], etc.)
model III	NICR ([19, 27], etc.), BGP-DA, ROFL [42]

(i.e., to follow the protocol) as they belong to the same administrative entity (or to multiple competing entities each providing the same service). In models II and III, however, one needs to consider strategic service nodes in addition to the strategic agents where the two sets could be the same. Some of the representative schemes in the literature that follow these service models are listed in Table 7.1.

In model (I) [$V_D \neq P$], there is a dedicated set of nodes V_D (possibly infrastructure) that keep the state about the identifier space while the players P reside on different nodes. DNS is one example of a centralized scheme that follows this model. In DNS, V_D is the set of root/gTLD (for global Top Level Domain) servers and the players are domain servers that keep zone files. Another scheme that uses this model and that is distributed is the recent DONA proposal [99] where V_D is the set of resolution handlers, and the players are generally objects on edge nodes. Another set of proposals that fits under this model is embodied by the Locator-ID-Split (LIS) work which aims at providing discoverability to edge sites (e.g., [66]) or nodes (e.g., [118]) in the Internet.

In model (II) [$V_D = P$], the state is kept on the same set of nodes that the players reside on. In such a model, the players themselves have a common interest in implementing the discovery scheme. The typical example here is Distributed Hash Tables (DHT).

In model (III) [$V_D = V = P$], the state is maintained on all the nodes V and the players are all the nodes. This model is common to proposals that perform routing on identifiers. One class of schemes that fits under this model is represented by Name Independent Compact Routing (NICR) [27]. In NICR, the forwarding function is aware of the node identifiers. BGP with de-aggregation is the another representative scheme here where the players or nodes are the ASs advertising the prefixes and where it is necessary for all nodes V to keep the state in order for prefix path discovery to succeed. Another recent scheme is the DHT-based ROFL [42], in which the routers are the nodes (if we ignore data objects here) that are themselves the players identified by flat identifiers (hashes). Note that models (II) and (III) are the same for our purposes and we shall not make the distinction between the two hereafter.

7.4 Incentives and Pricing

Having introduced the discovery problem and overviewed different discovery models used in the literature, we now proceed to motivate the need for incentives in discovery. Recall that in order for a node to be discoverable, a cost must be incurred by the set of service nodes V_D generally for maintaining state about the node's identifier. The term *state* in this context refers to the information stored on the service nodes to allow the players to be discoverable. The per-node state may be thought of as simply the node's routing table which is generally comprised of mappings from identifier to location information. The question that arises then is who pays for maintaining the state, and what incentive models are suitable for the different discovery models. In this section, we present our initial thoughts on incentive models that are applicable to each of the discovery models, and we set the stage for the BGP incentive model which is discussed in the next chapter.

Model I: $V_D \neq P$ Recall that in this setting, the players P are requesting a discovery service from a set of infrastructure service nodes V_D. When $V_D \neq P$, mechanism design and particularly Distributed Algorithmic Mechanism Design (DAMD) [60] in addition to general cost-sharing models [123] seem to be intuitive frameworks for modeling incentives and pricing. Different situations may arise based on whether the service nodes are obedient or not (obedient service nodes will not try to manipulate the protocol), belong to multiple competing economic entities or not, and on whether the mechanism is subsidized or not. Note that when the mechanism is subsidized, the designer of the mechanism does not have to worry about budget-balance where the latter means that the total payments made by the players must offset the total cost of providing the service.

Assume the service nodes to be obedient not competing, and consider the following DAMD model: each player has a valuation of being discoverable, which she presents to the mechanism. The mechanism logically controls (and operates on) the service nodes collecting all the players' valuations, the demand, and allocating payments back to the players to achieve a mix of goals. These goals could potentially include incentive-compatibility (or strategy-proofness), welfare maximization (or efficiency), and/or budget-balance. When the mechanism is subsidized, the goal of the mechanism is to maximize the social welfare (instead of budget-balance) under the constraint that a cost is associated with providing the service. In this sense, valuations of the service need to be declared truthfully by the players, and hence the goal is incentive-compatibility (especially when the mechanism is able to provide different levels of the service). A one-shot Vickery–Clarke–Groves (VCG) variant [123] is a natural solution here that could achieve efficiency and incentive-compatibility (again assuming that the mechanism could be subsidized in other ways). The VCG pricing scheme is a cost-sharing scheme, i.e., it shares the total cost of providing the service among the participating players. The mechanism always maximizes the social welfare of all the players and picks prices (cost shares) as follows: a player i pays an amount equal to the difference in the total welfare of the other players with and without player i's participation. This is equivalent to the damage caused by player i's participation.[2] Note that the budget-balance requirement becomes essential when the subsidization assumption does not hold since the total cost must be collected so that service nodes are paid for participating. For example, if a node is not compensated for the cost of keeping state about the rest of the network, the node will have no incentive to participate. It has been proven by Laffont and Green and later by Satterwaite's impossibility theorems [123] that cost-sharing mechanisms may be either strategy-proof and efficient, or strategy-proof and budget-balanced but not both.

When the service providers compete, then the one-shot mechanism design framework seems less practical. This case is more representative of model (I) than the no-competition case. The main idea here is that multiple competing Discovery Service Providers (DSPs) offer the service to the players. Each DSP is assumed to be owned and operated by an autonomous economic entity and DSPs compete for service or market share. Dynamic pricing models and repeated dynamics are more suitable here (see for example [22]).

Models II, III: $V_D = P$ When the set of service nodes $V_D = P$, players incur a cost due to participation of other players and the issue of incentives and pricing becomes even more challenging. In such a distributed setting, the traditional game theoretic and economic tools seem to be more applicable, since the centralized designer and the obedient service nodes assumptions inherent to the mechanism design framework are less practical. Consider BGP for example where every node that wishes to be discoverable introduces state about its identifier on every other node in the DFZ. NICR [19, 27] schemes on the other hand are less costly as they try to optimize the tradeoff between state and stretch (see Chaps. 5 and 6 for more on stretch and space/state tradeoffs in the context of compact routing). In this sense, a node that wishes to be discoverable must introduce state on a subset of other nodes in the network. In both examples above, one may directly recognize the incentive mismatch issue and the challenges inherent to the design of incentive and pricing models that are suitable for this setting. In the next chapter, we present one such model for BGP.

[2]This VCG pricing scheme is referred to as the Clark Pivot rule [123].

7.5 Conclusion

This chapter has motivated the need for considering strategic interactions in network design. We presented three games that highlight the most common solution concepts employed when analyzing strategic interactions among self-interested, utility-maximizing agents. The solution concepts reviewed are pure strategy Nash equilibrium (QUERY-GAME), dominant strategy equilibrium (ROUTING-GAME), and subgame perfect equilibrium (TRADE-GAME). The games are very relevant to the discussion of the next chapter (Chap. 8) where we present an incentive model for route distribution in the context of BGP. To illustrate the incentive issues that arise in discovery mechanism design, we presented a taxonomy of discovery schemes based on their business models. We highlighted two main models: one in which the set of service nodes is different from the set of players (model I), and another in which the two sets of nodes are the same (models II and III). We discussed our thoughts on suitable economic models for each of the two discovery models. The next chapter elaborates on the incentive issues that arise in the BGP scheme (BGP follows models II, III). We present an incentive model that allows for route distribution while aligning the incentives of all the participating agents.

Chapter 8
Participation Incentives in BGP

Nothing in the world is ever completely wrong. Even a stopped clock is right twice a day
Paulo Coelho, Brida

In this chapter, we use game theory to model a general participation game. The main problem we are interested in is how to achieve broad participation while aligning the incentives of all the participating agents. A consumer node is willing to invest some initial amount of money to get a set of networked nodes, alternatively agents, to participate in a desirable activity. The consumer, in this case a BGP speaker, desires to advertise itself; however, it may only communicate with its direct neighbors. Therefore, it must incentivize its neighbors to participate in further advertising its route, who then incentivize their neighbors to participate, and so on. We assume the commodity being traded to be the agent's participation. In the resulting game, agents choose their offers strategically and they are rewarded by volume of sales. We prove the existence of equilibria for specific utility functions and simple network structures.

8.1 Introduction

The Border Gateway Protocol (BGP) is a policy-based path vector protocol and is the de-facto protocol for Internet interdomain routing. The protocol's specification [132] was initially intended to empower domains with control over *route selection*, and *route propagation*. The commercialization of the Internet transformed Autonomous Systems (AS) into economic entities that act selfishly when implementing their internal policies and particularly the decisions that relate to route selection and propagation [43]. BGP is intrinsically about distributing route information to destinations (which are IP prefixes) to establish paths in the network. Path discovery, or simply discovery hereafter, starting with some destination prefix is the outcome of route distribution and route computation.

As discussed earlier in Chap. 7, accounting for and sharing the cost of discovery is an interesting problem whose absence from current path discovery schemes has

J.S. Khoury, C.T. Abdallah, *Internet Naming and Discovery*,
Signals and Communication Technology, DOI 10.1007/978-1-4471-4552-3_8,

led to critical economic and scalability concerns. As an example, the BGP control plane functionality is oblivious to cost. More clearly, a node (BGP speaker) that advertises a provider-independent prefix (identifier) does not pay for the cost of being discoverable. Such a cost, which may be large given that the prefix is maintained at every node in the Default-Free Zone (DFZ), is paid by the rest of the network. For example, Herrin [83] has preliminarily analyzed the non-trivial cost of maintaining a BGP route. Such incentive mismatch in the current BGP workings is further exacerbated by provider-independent addressing, multi-homing, and traffic engineering practices [115]. Given the fact that the number of BGP prefixes in the global routing table (or RIB) is constantly increasing at a rate of roughly 100,000 entries every 2 years [88], has motivated us to devise a model that accounts for distribution incentives in BGP.

A large body of work has focused on choosing the right incentives given that ASs are self-interested, utility-maximizing agents. While exploring incentives, most previous work has ignored the control plane incentives (route advertisement/distribution) and has instead focused on the forwarding plane incentives (e.g., transit costs).[1] One possible explanation for this situation is based on the following assumption: a node has an incentive to distribute routes to destinations since the node will get paid for transiting traffic to these destinations, and hence route distribution is ignored as it becomes an artifact of the transit process. We argue that this assumption is not economically viable by considering the arrival of a new customer (BGP speaker). While the servicing edge provider makes money from transiting the new customer's traffic to the customer, the middle providers do not necessarily make money while still incurring the cost to maintain and distribute the customer's route information. In this work, we separate the control plane incentives (incentives to distribute route information) from the forwarding plane incentives (incentives to forward packets) and use game theory to model a BGP distribution game. The main problem we are interested in is how to allow BGP prefix information to be distributed globally while aligning the incentives of all the participating agents.

8.1.1 A Simple Distribution Model

We synthesize many of the ideas and results from [59, 75, 97, 108] into a coherent model for studying BGP route distribution incentives. Influenced by the social network query propagation model of Kleinberg and Raghavan [97], we use a completely distributed model in the sense that it does not assume a central bank (in contrast to previous work on truthful mechanisms [123]).

A destination d advertises its prefix and wishes to invest some initial amount of money r_d in order to be globally discoverable (or so that the information about d

[1] In this chapter, we use the term "control plan" to refer only to route prefix advertisements (not route updates) as we assume that the network structure is static.

is globally distributed). Since d may distribute its information to its direct neighbors only, d needs to provide incentives to get the information to propagate deeper into the network. Therefore, d must incentivize its neighbors to be distributors of its prefix/route, who then incentivize their neighbors to be distributors, and so on. A distributor node is rewarded based on the role it plays in the outcome routing tree to d, T_d. The utility of the transit node i from distributing d's route, as we shall describe shortly, increases with the number of nodes that route to d through i, hence the incentive to distribute. While we take BGP as the motivating application, we are interested in the general setting of incentivizing a set of agents to participate in some activity that is of interest to a consumer node. For example, the consumer node might want to measure a signal, to solve a problem, or to distribute a good. The consumer receives a positive utility from every agent that participates, whereas an agent incurs a cost to participate. The agents are located on a network and communication, in the form of trade, may only occur between directly connected agents. Other general instances of the problem include: (1) advertisement and/or search in social networks where an advertiser wishes to broadly advertise a piece of information within a social network or to get a query answered, while only having access to a small set of source agents [86, 97]; or (2) group problem solving as demonstrated with the DARPA network challenge [4]. In an effort to explore the power of social networking in team problem solving, DARPA announced a network challenge to find the locations of 10 balloons dispersed across the continental United States in a timely manner, a problem that requires participation incentives.

Our model seems to correctly capture many of the details behind how policy-based BGP (and in general path-vector protocols) works and the inherent incentives required. Additionally, the model is consistent with the simple path vector formulation introduced by Griffin in [75]. More clearly, it is widely accepted that each AS participating in BGP has the following decisions to make as part of its decision space:

- import policy: a decision on which routes to d to consider,
- route selection: a decision on what route to d to choose among the multiple possible routes,
- export policy: a decision on who to forward the advertisement to among its direct neighbors.

All three policies are captured in the game model we describe next.

There are two main properties of interest when it comes to the BGP game model: *convergence*, and *incentives*. The BGP interdomain routing protocol handles complex interactions between autonomous, competing economic entities that can express local preferences over the different routes. Given the asynchronous interactions among the ASs and the partial information, convergence of BGP to a stable solution becomes an essential property to aim for when studying policies. Griffin et al. [75] defined the stable paths problem which is widely accepted as the general problem that BGP is solving. The authors formulated a general sufficient condition under which the protocol converges to an equilibrium state, mainly the "no dispute wheels" condition. A game-theoretic model was recently developed by Levin et al. [108] that builds on the stable paths formalization and studies the incentive-compatibility question. In addition to convergence, incentive issues are crucial to the success and stability of BGP mainly since nodes are assumed to be selfish entities that act strategically to optimize their utility. In this sense, any distribution and route computation mechanism or policy benefits from aligning the incentives of the players to achieve the mechanism's goals [58, 59, 108, 123].

8.1.2 Results

In Sect. 8.2, we present the general distribution game. In this game, a player's pure strategy involves deciding on a "best" route to d as well as determining the reward to offer to her direct neighbors. We define the player's utility as a function of the volume of downstream players that she can recruit. Based on the assumption that the advertiser or destination, d, receives a fixed marginal utility from each player that maintains a route to d. In this chapter, we are mainly interested in studying the existence of equilibria in the general game. Our main results include:

- First, in Sect. 8.3 we prove that in order to maximize her utility, a player always chooses the route with the highest promised reward. We refer to the resulting policy as the Highest Rewards Path (HRP). We show that under HRP, the BGP protocol always converges to a stable routing tree for any strategy profile by employing the sufficient condition for convergence defined by Griffin et al. [75]. The convergence result allows us to focus on the existence of equilibria as it directly means that any equilibrium strategy profile converges under the BGP protocol to a stable tree.
- Due to the complexity of the strategic dependencies and the competition dynamics, Sect. 8.4 presents the initial equilibrium results on the simplest possible class of graphs with and without competition. Particularly, we present existence results for: (1) the line (and the tree) graphs which involve no competition, and (2) the ring graph which involves competition. We show that a subgame-perfect equilibrium always exists for the game induced on the line graph and on the tree. However, no such equilibrium exists for the game induced on the ring graph (with an even number of nodes) due to oscillation of *best-response* dynamics under competition when the incentive r_d is "large". To the best of our knowledge, this is the first result to consider competition which has not been studied in similar previous work [97, 109]. While the full game does not have a subgame-perfect equilibrium, we show that there always exists a Nash equilibrium for a special class of subgames. This requires us to first quantify the growth of rewards in the game, or in other words the minimum incentive r_d such that there exists an equilibrium outcome which is a spanning tree.
- Finally, Sect. 8.4.5 extends the static version of the game to a repeated version. We show how a Nash equilibrium may be constructed in a finitely repeated version of the game by adding a *convergence* constraint on utility which essentially disincentivizes oscillation.

8.2 The General Game

Reusing notation from [59, 108], we consider a graph $G = (V, E)$ where V is a set of n nodes (alternatively termed players, or agents) each identified by a unique index $i = \{1, \ldots, n\}$, and a destination d, and E is the set of edges or links. Without loss of generality, we study the BGP discovery/route distribution problem for some fixed destination AS with prefix d (as in [59, 75, 108]). The model is extendable to all possible destinations (BGP speakers) by noticing that route distribution and computation are performed independently per prefix. The destination d is referred to as the *advertiser* and the set of players in the network are termed *seekers*. Seekers may be distributors who participate in distributing d's route information to other seeker nodes or consumers who simply consume the route (leaf nodes in the outcome distribution tree). For each seeker node j, let $P(j)$ be the set of all routes to d that are known to j through advertisements, $P(j) \subseteq \mathcal{P}(j)$, the latter being the set of all simple routes from j. The empty route $\phi \in \mathcal{P}(j)$. Denote by $R_j \in P(j)$ a simple route from j to the destination d, with $R_j = \phi$ when no route exists at j. Let $(k, j)R_j$ be the route formed by concatenating link (k, j) with R_j, where $(k, j) \in E$. Denote by $B(i)$ the set of direct neighbors of node i and let $next(R_i)$ be the next hop node on the route R_i from i to d. Define node j to be an *upstream* node relative to node i when $j \in R_i$. The opposite holds for a *downstream* node. Finally, we use $r_{\text{next}(R_i)}$ to refer to the reward that the upstream parent from i on R_i offers to i. For example in Fig. 8.1, $next(R_5) = 3$ and 3 is an upstream node relative to 5.

The general distribution game is as follows: destination d first exports its prefix information to its neighbors promising them a reward $r_d \in \mathbb{Z}^+$ ($r_d = 10$ in Fig. 8.1) which directly depends on d's utility of being discoverable. A node j (a player) in turn receives offers from its neighbors where each neighbor i's offer takes the form of a reward r_{ij}. A reward r_{ij} that a node i offers to some direct neighbor $j \in B(i)$ is a contract stating that i will pay j an amount that is a function of r_{ij} and of the set of downstream nodes k that decide to route to d through j (i.e., $j \in R_k$ and $R_j = (j, i)R_i$). Note that such a reward model requires that the downstream nodes k notify j of their best route so that the latter can claim its reward from its upstream parent. After receiving the offers, player j strategizes by selecting a route among the possibly multiple advertised routes to d, say $(j, i)R_i$, and deciding on a reward $r_{jl} < r_{ij}$ to send to each *candidate* neighbor $l \in B(j)$ that it has not received a competing offer from. Note then that $r_{lj} < r_{jl}$ where $r_{lj} = 0$ means that j did not receive an offer from neighbor l. Node j then pockets the difference $r_{ij} - r_{jl}$. The process repeats up to some depth that is directly dependent on the initial investment r_d as well as on the strategies of the players. We intentionally keep this reward model abstract at this point, but will revisit it later in the discussion when we define more specific utility functions. For example, in Fig. 8.1, node d promises its neighbor set $\{1, 2\}$ a reward $r_d = 10$. Node 1 exports route $(1, d)$ to its neighbor promising a reward $r_{13} = 8$. Similarly node 2 exports the route $(2, d)$ to its neighbor set $\{3, 4\}$ with $r_{23} = r_{24} = 7$ and so on. Clearly in this model, we assume that a player can strategize per neighbor, presenting different rewards to different neighbors. This assumption is based on the autonomous nature of the nodes and the current practice

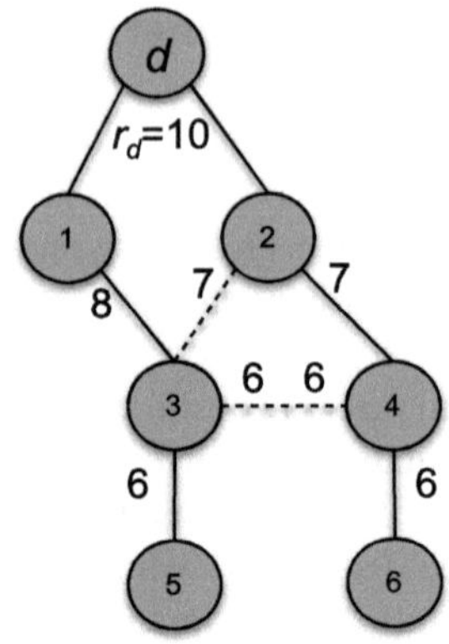

Fig. 8.1 Sample network (not at equilibrium): *Solid lines* indicate an outcome tree T_d under the advertised rewards

in BGP where policies may differ significantly across neighbors (as with the widely accepted Gao–Rexford policies [67] for example).

Assumptions To keep our model tractable, we make several simplifying assumptions. In particular, we assume that:

1. the graph is at steady state for the duration of the game, i.e., we do not consider topology dynamics;
2. the advertiser d does not differentiate among the different players (ASs) in the network, i.e., the ASs are indistinguishable to d;
3. the advertised rewards are integers and are strictly decreasing with depth, i.e., $r_{ij} \in \mathbb{Z}^+$ and $r_{ij} < r_{\text{next}(R_i)}$, $\forall i, j$. We let 1 unit be the cost of distribution (a similar assumption was taken in [97] to avoid the degenerate case of never running out of rewards, referred to as "Zeno's Paradox");
4. a node that does not participate gets zero utility;
5. finally, our choice of the utility function isolates a class of policies which we refer to as the Highest Reward Path (HRP). As the name suggests, HRP policies incentivize players to choose the path that promises the highest reward. Such class of policies may be defined more generally to account for more complex cost structures as part of the decision space.[2] We assume for the scope of this work that transit costs are extraneous to the model. This is a restrictive assumption given that BGP allows for arbitrary and complex policies that are generally modeled with a valuation or preference function over the different routes to d (see [59, 75]).

Strategy Space Given a set of advertised routes $P(i)$ where each route $R_i \in P(i)$ is associated with a promised reward $r_{\text{next}(R_i)} \in \mathbb{Z}^+$, a *pure strategy* $s_i \in S_i$ of an autonomous node i comprises two decisions:

[2]Metric-based policies could be modeled with HRP by fixing one of the players' decisions. For example, fixing $r_{ij} = r_{\text{next}(R_i)} - 1$, $\forall i, j$ results in hop count metric; or alternatively setting $r_{ij} = r_{\text{next}(R_i)} - c_i$, where c_i is some local cost to the node results in Least Cost Path (LCP) policy [59], etc.

- After receiving offers from neighboring nodes, choose a single "best" route $R_i \in P(i)$ (where "best" is defined shortly in Theorem 8.1);
- Choose a reward vector $r_i = [r_{ij}]_j$ promising a reward r_{ij} to each candidate neighbor j (and export route and reward to the respective candidate neighbors).

A strategy profile $\mathbf{s} = (s_1, \dots, s_n)$ and a reward r_d define an outcome of the game.[3] Every outcome determines a set of paths to destination d given by $O_d = (R_1, \dots, R_n)$. A utility function $u_i(\mathbf{s})$ for player i associates every outcome with a real value in $\mathbb{R}$. We use the notation s_{-i} to refer to the strategy profile of all players excluding i. The Nash equilibrium is defined as follows:

Definition 8.1 A Nash Equilibrium (NE) is a strategy profile $\mathbf{s}^* = (s_1^*, \dots, s_n^*)$ such that no player may deviate profitably from her strategy given everyone else's equilibrium strategy, i.e., for each player i, $u_i(s_i^*, s_{-i}^*) \geq u_i(s_i, s_{-i}^*)$, $\forall s_i \in S_i$.

Cost The cost of participation is local to the node and includes for example the cost associated with the effort that a node spends in maintaining the route information.[4] Other cost factors that depend on the volume of traffic (proportional to the number of downstream nodes in the outcome O_d) are more relevant to the forwarding plane and as mentioned earlier in the assumptions, we ignore this cost in the current model. Hence, we simply assume that every player i incurs a cost c_i which is the cost of participating. We assume for the scope of this chapter that the local cost is constant with $c_i = c = 1$.

Utility We experiment with a simple class of utility functions which rewards a node linearly based on the number of sales that the node makes. This model incentivizes distribution and potentially requires a large initial investment from d. More clearly, define $N_i(\mathbf{s}) = \{j \in V \backslash \{i\} \mid i \in R_j\}$ to be the set of nodes that choose their best route to d going through i (nodes downstream of i) and let $\delta_i(\mathbf{s}) = |N_i(\mathbf{s})|$. Let the utility of a node i from an outcome or strategy profile $\mathbf{s}$ be:

$$u_i(\mathbf{s}) = (r_{\text{next}(R_i)} - c_i) + \sum_{\{j \mid i = \text{next}(R_j)\}} (r_{\text{next}(R_i)} - r_{ij})(\delta_j(\mathbf{s}) + 1). \tag{8.1}$$

The first term $(r_{\text{next}(R_i)} - c_i)$ of Eq. (8.1) is incurred by every participating node and is the one unit of reward from the upstream parent on the chosen best path minus the local cost. Based on the fixed cost assumption, we often drop this first term when comparing player payoffs from different strategies since the term is always

---

[3] We abuse notation hereafter and we refer to the outcome with simply the strategy profile $\mathbf{s}$ where it should be clear from context that an outcome is defined by the tuple $\langle \mathbf{s}, r_d \rangle$. Notice that a strategy profile may be associated with an outcome if we model r_d as an action. We refrain from doing so to make it explicit that r_d is not strategic.

[4] A preliminary estimate of this cost is shown by Herrin [83] to be $0.04 per route/router/year for a total cost of at least $6,200 per year for each advertised route assuming there are around 150,000 DFZ routers that need to be updated.

positive when $c = 1$. The second term of Eq. (8.1) (the summation) is incurred only by distributors and is the total profit made by i where $(r_{\text{next}(R_i)} - r_{ij})(\delta_j(\mathbf{s}) + 1)$ is i's profit from the sale to neighbor j (which depends on δ_j). We assume here that node i gets no utility from an oscillating route and gets positive utility when R_i is stable. A rational selfish node will always try to maximize its utility when choosing $s_i = (R_i, [r_{ij}]_j)$. There is an inherent tradeoff between $(r_{\text{next}(R_i)} - r_{ij})$ and $(\delta_j(\mathbf{s}))$ s.t. $i = next(R_j)$ when trying to maximize the utility in Eq. (8.1) in the face of competition as shall become clear later. A higher promised reward r_{ij} allows the node to compete (and possibly increase δ_j) but will cut the profit margin. Finally, we implicitly assume that the destination node d gets a constant marginal utility of r_d for each distinct player that maintains a route to d—the marginal utility of being discoverable by any seeker—and declares r_d truthfully to its direct neighbors (i.e., r_d is not strategic).

8.3 Convergence Under HRP

Before proceeding with the game model, we first prove the following theorem, which results in the Highest Reward Path (HRP) policy.

Theorem 8.1 *In order to maximize its utility, node i must always choose the route R_i with the highest promised reward, i.e., such that $r_{\text{next}(R_i)} \geq r_{\text{next}(R_l)}$, $\forall R_l \in P(i)$.*

Proof The case for $|B(i)| = 1$ is trivial. The case for $|B(i)| = 2$ is trivial as well since i will not be able to make a sale to the higher reward neighbor by choosing the lower reward offer. Assume that node i has more than 2 neighbors and that any two neighbors, say k, l advertise routes $R_k, R_l \in P(i)$ s.t. $k = next(R_k), l = next(R_l)$ and $r_{ki} < r_{li}$, and assume that i's utility for choosing route R_k over R_l either increases or remains the same, i.e., $u_i^{R_k} \geq u_i^{R_l}$. We will show by contradiction that neither of these two scenarios could happen.

Scenario 1: $u_i^{R_k} > u_i^{R_l}$ From Eq. (8.1), it must be the case that either (case 1) node i was able to make at least one more sale to some neighbor j who would otherwise not buy, or (case 2) some neighbor j who chooses $(j,i)R_i$ can strictly increase her $\delta_j(\mathbf{s})$ when i chooses the lower reward path R_k. For case 1, and assuming that r_{ij} is the same when i chooses either route, it is simple to show that we arrive at a contradiction in the case when $j \in \{k, l\}$ (mainly due to the strictly decreasing reward assumption, i.e., $r_i < r_{\text{next}(.)}$); and in the case when $j \notin \{k, l\}$, it must be the case that j's utility increases with i's route choice, i.e., $u_j^{(j,i)R_k} > u_j^{(j,i)R_l}$. This contradicts with Eq. (8.1) since w.r.t. j, both routes have the same next hop node i. The same analogy holds for case 2.

Scenario 2: $u_i^{R_k} = u_i^{R_l}$ Using the same analogy of scenario 1, there must exist at least one neighbor j of i that would buy i's offer only when the latter chooses R_k, or otherwise node i is able to strictly increase its utility by choosing R_l pocketing more profit. □

The theorem implies that a player could perform her two actions sequentially, by first choosing the highest reward route R_i, then deciding on the reward vector to export to its neighbors. Thus, we shall represent player i's strategy hereafter simply with the rewards vector $[r_{ij}]$ and it should be clear that player i always chooses the "best" route to be the route with the highest promised reward. For the scope of this work, when the rewards are equal, we assume that a node breaks ties consistently. We refer the interested reader to a parallel result that develops the random tie-breaking assumption instead [93].

The question we attempt to answer here is whether the BGP protocol dynamics converge to a unique outcome tree T_d under some strategy profile $\mathbf{s}$. A standard model for studying the convergence of BGP protocol dynamics was introduced by Griffin et al. [75], and assumes BGP is an infinite round game in which a *scheduler* entity decides on the *schedule*, i.e., which players participate at each round (models the asynchronous operation of BGP). The authors devised the "no dispute wheels" condition [75], which is the most general condition known to guarantee convergence of possibly "conflicting" BGP policies to a unique stable solution (tree). From Theorem 8.1, it may be easily shown that "no dispute wheels" exist under HRP policy, i.e., when the nodes choose highest reward path breaking ties consistently. This holds since any dispute wheel violates the assumption of strictly decreasing rewards on the reward structure induced by the wheel. Hence, the BGP outcome converges to a unique tree T_d [75] under any strategy profile $\mathbf{s}$. This result allows us to focus on the existence of equilibria as it directly means that the BGP protocol dynamics converges to a tree under any equilibrium strategy profile.

8.4 Equilibria

In the general game model defined thus far, the tie-breaking preferences of the players is a defining property of the game, and every outcome (including the equilibrium) depends on the initial reward/utility r_d of the advertiser. Studying the equilibria of the general game for different classes of utility functions and for different underlying graph structures is not an easy problem due to the complexity of the strategic dependencies and the competition dynamics. Hence, we start by studying the game on the simplest possible class of graphs with and without competition. We assume *full information* as we are interested in studying the existence question initially rather than how the players would arrive at the equilibrium.[5] Particularly, we present existence results for the simplest two classes of graphs: (1) the line (and the tree) graphs which involve no competition, and (2) the ring graph which involves competition.

[5]This of course is an interesting question in its own right.

To study the existence of equilibria on the simple line and ring graphs, we fix the *schedule* of play (i.e., who plays when?) as we formalize shortly. We start by examining a static version of the full-information game in which each player plays once at a single stage based on proximity to d, and we then proceed to examine the repeated version of the static game.

8.4.1 The Static Multi-Stage Game with Fixed Schedule

In order to apply the correct solution concept, we fix the *schedule* of play. The schedule is based on the inherent order of play in the model: recall that the advertiser d starts by advertising itself and promising a reward r_d; the game starts at stage 1 where the direct neighbors of d, i.e., the nodes at distance 1 from d, observe r_d and play simultaneously by choosing their rewards while the rest of the nodes "do-nothing". At stage 2, nodes at distance 2 from d observe the stage 1 strategies and then play simultaneously and so on. Stages in this *multi-stage game with observed actions* [65] have no temporal semantics. Rather, they identify the network positions which have strategic significance. The closer a node is to the advertiser, the more power such a node has due to the strictly decreasing rewards assumption. The key concept here is that it is the *information sets* [65] that matter rather than the time of play, i.e., since all the nodes at distance 1 from d observe r_d before playing, all these nodes belong to the same information set whether they play at the same time or at different time instants. We refer to a single play of the multi-stage game as the *static* game. We resort to the multi-stage model (the fixed schedule) on our simple graphs to eliminate the synchronization problems inherent in the BGP protocol and to focus instead on the existence of equilibria. By restricting the analysis to the fixed schedule, we do not miss any equilibria. This is due to the fact that the fixed schedule is only meant to replace the notion of "fair and infinite schedule" [75] with a more concrete order of play. The resulting game always converges in a single play for any strategy profile, and the outcome tree is necessarily one of shortest-paths (in terms of number of hops).[6] The main limitation of this model, however, is that it cannot deal with variable costs c_i for which the outcome (HRP tree) might not be a shortest-path tree.

Formally, and using notation from [65], each player i plays only once at stage $k > 0$ where k is the distance from i to d in number of hops. At every other stage, the player plays the "do nothing" action. The set of player actions at stage k is the stage-k action profile, denoted by $a^k = (a_1^k, \dots, a_n^k)$. Further, denote by $h^{k+1} = (r_d, a^1, \dots, a^k)$, the *history* at the end of stage k which is simply the initial reward r_d concatenated with the sequence of actions at all previous stages. We let $h^1 = (r_d)$.

[6]This follows in the multi-stage game since a player at stage k will not offer rewards to its neighbors at stage $l < k$, i.e., rewards flow in one direction away from d. The outcome is necessarily a shortest-path tree since every player at stage k must choose its best route from the offers its received from neighbors at stage $k - 1$.

Finally, $h^{k+1} \subset H^{k+1}$ the latter being the set of all possible stage-k histories. When the game has a finite number of stages, say $K+1$, then a terminal history h^{K+1} is equivalent to an outcome of the game (which is a tree T_d) and the set of all outcomes is H^{K+1}.

The pure strategy of player i who plays at stage $k > 0$ is a function of the history and is given by $s_i : H^k \to \mathbb{R}^{m_i}$ where m_i is the number of direct neighbors of player i that are at stage $k+1$ (implicit here is that a player always chooses the highest reward route). Starting with r_d (which is h^1), it is clear how the game produces actions at every later stage based on the player strategies resulting in a terminal action profile or outcome. Hence, given r_d, an outcome in H^{K+1} may be associated with every strategy profile $\mathbf{s}$, and so the definition of Nash equilibrium (Definition 8.1)) remains unchanged. Finally, it is worthwhile noting that the "observed actions" requirement (where a player observes the full history before playing) is not necessary for our results in the static game as we shall see in the construction of the equilibrium strategies. Keeping this requirement in the model allows us to classify the play from some stage onward, contingent on a history being reached as a subgame in its own right as we describe next.

Definition 8.2 [65] A *proper subgame* of a full game is a restriction of the full game to a particular history. The subgame inherits the properties of the full game such as payoffs and strategies while simply restricting those to the history.

In our game, each stage begins a new subgame, which restricts the full game to a particular history. For example, a history h^k begins a subgame $G(h^k)$ such that the histories in the subgame are restricted to $h^{k+1} = (h^k, a^k)$, $h^{k+2} = (h^k, a^k, a^{k+1})$, and so on.

Definition 8.3 [65] A strategy profile $\mathbf{s}^* = (s_1^*, \ldots, s_n^*)$ is a *subgame-perfect equilibrium* if it is a Nash equilibrium for every proper subgame of the full game.

Hereafter, the general notion of equilibrium we use is the Nash equilibrium and we shall make it clear when we generalize to subgame-perfect equilibria. We are only interested in pure-strategy equilibria [65] and in studying the existence question as the incentive r_d varies. A Nash equilibrium hereafter is a pure-strategy Nash equilibrium. We now proceed to study the equilibria on special networks.

No Competition: The Line Graph and the Tree In the same spirit as [97] we inductively construct the equilibrium for the line graph (simply referred to as the line hereafter) of Fig. 8.2 given the utility function of Eq. (8.1). We present the result for the line which may be directly extended to trees. Before proceeding with the construction, notice that for the line, $m_i = 1$ for all players except the leaf player since each of those players has a single downstream neighbor. In addition, $\delta_i(\mathbf{s}) = \delta_j(\mathbf{s}) + 1, \forall i, j$ where j is i's child ($\delta_i = 0$ when i is a leaf). We shall refer to both the player and the stage using the same index since our intention should be clear from the context. For example, the child of player i is $i+1$ and its parent is $i-1$

Fig. 8.2 Line graph: a node's index is the stage at which the node plays; d advertises at stage 0; $K = n$

where player i is the player at stage i. Additionally, we simply represent the history $h^{k+1} = (r_k)$ for $k > 0$ where r_k is the reward promised by player k (player k's action). The strategy of player k is therefore $s_k(h^k) = s_k(r_{k-1})$ which is a singleton (instead of a vector) since $m_i = 1$ (for completeness, let $r_0 = r_d$). This is a *perfect information* game [65] since a single player moves at each stage and has complete information about the actions of all players at previous stages. Hence, backward induction may be used to construct the subgame-perfect equilibrium.

We construct the equilibrium strategy s^* inductively as follows: first, for all players i, let $s_i^*(x) = 0$ when $x \leq c$ (where c is assumed to be 1). Then assume that $s_i^*(x)$ is defined for all $x < r$ and for all i. Obviously, with this information, every player i may compute $\delta_i(x, s_{-i}^*)$ for all $x < r$. This is simply due to the fact that δ_i depends on the downstream players from i who must play an action or reward strictly less than r. Finally, for all players i we let $s_i^*(r) = \arg\max_x (r - x)\delta_i(x, s_{-i}^*)$ where $x < r$.

Theorem 8.2 *The strategy profile s^* is a subgame-perfect equilibrium.*

Sketch of Proof The proof for the line is straightforward and follows from backward induction by constructing the optimal strategies starting with the last player (player K) first, then the next-to-last, and so on up to player 1. The strategies are optimal for every history (by construction) and given the utility function defined in Eq. (8.1), no player can move profitably. Notice that in general when $r_{\text{next}(R_i)} \leq c$, propagation of the reward will stop simply because at equilibrium no player will want a negative utility and will prefer to not participate instead (the case with the leaf player). □

The proof may be directly extended to the tree since each player in the tree has a single upstream parent as well and backward induction follows in the same way. On

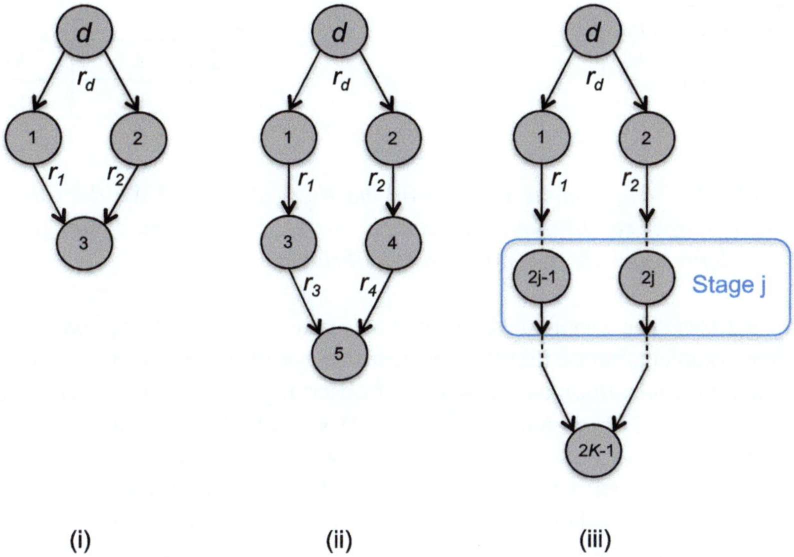

Fig. 8.3 Ring network with even number of nodes: (**i**) 2-stage game, (**ii**) 3-stage game, and general (**iii**) K-stage game

the tree, the strategies of the players that play simultaneously at each stage are also independent.

Competition: The Ring As opposed to the line, we present next a negative result for the ring graph (simply referred to as the ring hereafter). In a ring, each player has a degree of 2 and $m_i = 1$ again for all players except the leaf player. We consider rings with an even number of nodes due to the direct competition dynamics. Figure 8.3 shows the 2-stage, the 3-stage, and general K-stage versions of the game. In the multi-stage game, after observing r_d, players 1 and 2 play simultaneously at stage 1 promising rewards r_1 and r_2, respectively, to their downstream children, and so on. We shall refer to the players at stage j using ids $2j-1$ and $2j$ where the stage of a player i, denoted as $l(i)$, may be computed from the id as $l(i) = \lceil \frac{i}{2} \rceil$. For the rest of the discussion, we assume WLOG that the player at stage K (with id $2K-1$) breaks ties by choosing the route through the left parent $2K-3$ (consistent tie-breaking).

For the 2-stage game in Fig. 8.3, it is easy to show that an equilibrium always exists in which $s_1^*(r_d) = s_2^*(r_d) = (r_d - 1)$ when $r_d > 1$ and 0 otherwise. This means that player 3 enjoys the benefits of *perfect competition* due to the Bertrand-style competition [65] between players 1 and 2. The equilibrium in this game is independent of player 3's preference for breaking ties. We now present the following negative result.

Fig. 8.4 The payoff matrix of players 1 and 2 for the 3-stage game on the ring of Fig. 8.3(ii) when $r_d = 6$

r_1 \ r_2	1	3
2	8,5	4,6
3	6,6	3,3

Claim 8.1 *The 3-stage game induced on the ring (of Fig.* 8.3(ii)) *does not have a subgame-perfect equilibrium. Particularly, there exists a class of subgames for* $h^1 = rd > 5$ *for which there is no Nash equilibrium.*

Sketch of Proof The proof makes use of a counterexample. Using the backward induction argument, notice first that the best strategy of players 3 and 4 is to play a *Bertrand-style competition* as follows: after observing $a^1 = (r_1, r_2)$, player 3 plays $r_3 = 0$ when $r_1 = 1$, $r_3 = \min(r_1 - 1, r_2 - 1)$ when both $r_1 > 1$ and $r_2 > 1$, and $r_3 = 1$ when $r_1 > 1$ and $r_2 = 1$. Player 4 plays symmetrically. Knowing that, players 1 and 2 will choose their strategies simultaneously and no equilibria exist for $r_d > 5$ due to oscillation of the best-response dynamics. This may be shown by examining the strategic form game, in normal/matrix form, between players 1 and 2 (in which the utilities are expressed in terms of r_d). We briefly show the subgame for $r_d = 6$ and we leave the elaborate proof as an exercise for the interested reader. Figure 8.4 shows the payoff matrix of players 1 and 2 for playing actions $r_1 \in \{2, 3\}$ (rows) and $r_2 \in \{1, 3\}$ (columns), respectively. The payoff shown is taken to be $u_i = (r_d - r_{ij})\delta_i$ ignoring the first term of Eq. (8.1). The actions shown are the only remaining actions after applying iterated strict dominance, i.e., all other possible actions for the players are strictly dominated. Clearly, no pure-strategy Nash equilibria exist.[7] The argument could be directly extended to any $r_d > 5$ since player 2 will still have the incentive to oscillate. □

The value $r_d > 5$ signifies the breaking point of equilibrium or the reward at which player 2, when maximizing her utility $(r_d - r_2)\delta_2$, will always oscillate between competing for 5 (by playing large r_2) or not (by playing small r_2). We elaborate on this value later in Sect. 8.4.4. Hence, under the linear utility given in Eq. (8.1), an equilibrium does not exist on the simple ring. This negative result for the game induced on the 3-stage ring may be directly extended to the general game for the K-stage ring by observing that a class of subgames $G(h^{K-2})$ of the general K-stage game are identical to the 3-stage game. While the full game does not have an equilibrium for $K > 2$ stages, we shall show next that there always exists an equilibrium for the special subgame $G(r_d^*)$ (for $h^1 = r_d^*$), where the reward r_d^* is the minimum incentive to guarantee that d's route is globally distributed at equilibrium. We define and compute r_d^* next before constructing the equilibrium.

[7] There is, however, a single mixed strategy equilibrium in which player 1 plays $r_1 = 2$ with probability $\frac{2}{3}$ while player 2 plays $r_2 = 1$ with probability $\frac{1}{2}$, yielding expected payoffs 6 and 5 for players 1 and 2, respectively.

8.4.2 Growth of Incentives

We answer the following question: find the minimum incentive r_d^*, as a function of the depth of the network K (equivalently the number of stages in the multi-stage game), such that there exists an equilibrium outcome for the subgame $G(r_d^*)$ that is a spanning tree. We seek to compute the function f such that $r_d^* = f(K)$. First, we present a result for the line, before extending it to the ring. On the line, K is simply the number of players, i.e., $K = n$.

Lemma 8.1 *On the line graph, we have* $f(0) = 0$, $f(1) = 1$, $f(2) = 2$, *and* $\forall k > 2$

$$f(k) = (k-1)f(k-1) - (k-2)f(k-2). \tag{8.2}$$

Proof First, $f(0) = 0$, $f(1) = 1$ and $f(2) = 2$ are trivially true given the utility function of Eq. (8.1). The proof uses induction on the depth of the network. First, for the base case $k = 3$, in the 3-stage line the Nash equilibrium is for player 1, the player at distance 1 from d, to play $r_1 = 2$ and for player 2 to play $r_2 = 1$ (in every NE, $s_i(1) = 0$, $\forall i$). Given $r_d^* = f(3) = 3$, the utility of player 1 is $u_1 = (3-2)2 \geq (3 - r_2')\delta_2'$, $\forall r_2' < 3$. Similarly player 2 may not move profitably from playing $r_2 = 1$.

Assume $f(x) = (x-1)f(x-1) - (x-2)f(x-2)$ holds $\forall x < k$. We construct the k-stage game from the $(k-1)$-stage game by adding a node/player between node d and node 1 in the $(k-1)$-stage game. Notice the player 2 in the k-stage game used to be player 1 in the $(k-1)$-stage game. By definition of f, in the k-stage game, when player 1 plays $r_1 = f(k-1)$ then $\delta_1 = (k-1)$ and no player $i, 2 \leq i \leq k$ may deviate profitably from playing $r_i = f(k-i)$. Here $r_1 = f(k-1)$ is the minimum reward to get a $\delta_1 = (k-1)$. In general, it holds by construction of f that there are k possible outcomes for player 1, corresponding to the values $\delta_1 = 0, 1, \ldots, k-1$. For each of these outcomes, we have an action for player 1, $r_1 = f(x)$, which results in the outcome tree corresponding to $\delta_1 = x$, $\forall x < k$ and such that no player besides player 1 may deviate profitably contingent on player 1 playing $r_1 = f(x)$. (In this outcome player i plays $f(x - i + 1)$ $\forall 2 \leq i \leq n$.) In order for $\delta_1 = k - 1$ to be the equilibrium outcome, it must be the case that $r_1 = f(k-1)$ maximizes player 1's utility given r_d (and hence no player including player 1 may deviate profitably), i.e., it must be that $\forall 2 \leq j \leq k$

$$\big(r_d - f(k-1)\big)(k-1) \geq \big(r_d - f(k-j)\big)(k-j).$$

This condition is equivalent to

$$\big(r_d - f(k-1)\big)(k-1) \geq \big(r_d - f(k-2)\big)(k-2) \tag{8.3}$$

since $(r_d - f(k-2))(k-2) \geq (r_d - f(k-j))(k-j)$, $\forall 3 \leq j \leq k$ and for $r_d \geq f(k-1)$. Equation (8.3) implies that $r_d \geq (k-1)f(k-1) - (k-2)f(k-2)$. The minimum such incentive is

$$r_d^* = f(k) = (k-1)f(k-1) - (k-2)f(k-2) \tag{8.4}$$

which is greater than $f(k-1)$ concluding the proof. □

Notice that $f(K)$ grows exponentially with the depth K of the line network.[8] By subtracting $f(k-1)$ from both sides of the recurrence relation, it may be shown that

$$f(k) - f(k-1) = (k-2)!. \tag{8.5}$$

8.4.3 A Special Subgame

We now revisit the K-stage game of Fig. 8.3(iii) on the ring and we focus on a specific subgame which is the restriction of the full game to $h_1 = r_d^* = f(K)$, and we denote this subgame by $G(r_d^*)$. Consider the following strategy profile $\mathbf{s}^*$ for the subgame: players at stage j play $s_{2j-1}^*(h^j) = f(K-j)$, and $s_{2j}^*(h^j) = f(K-j-1)$, $\forall 1 \leq j \leq K-1$, and let $s_{2K-1}^*(h^K) = 0$.

Theorem 8.3 *The profile* $\mathbf{s}^*$ *is a Nash equilibrium for the subgame* $G(r_d^*)$ *on the* K*-stage ring,* $\forall K > 2$.

Proof Notice first that the complete history h^{K+1} which corresponds to r_d^* and $\mathbf{s}^*$ is an outcome that is a spanning tree (each player chooses the best route through the upstream parent while the last player $2K-1$ prefers the left parent who is promising a higher reward). We will show that no player i can deviate from playing s_i^* given s_{-i}^* by considering the players at each stage j, $\forall 2 \leq j \leq K-1$ first and then we extend the reasoning to the players at stage 1. For the players at stage j we show that player $2j-1$ may not deviate profitably from playing $s_{2j-1}^*(h^j) = r_{2j-1} = f(K-j)$ given the strategies of the rest of the players (particularly given $s_{2j}^*(h^j) = r_{2j} = f(K-j-1)$), and the same for player $2j$. Given that $r_{2j} < r_{2j-1}$ (i.e., player $2j$ not competing with player $2j-1$), then by definition of the function f, there exists an outcome on the ring such that $\delta_{2j-1} = K-j$ when $r_{2j-1} = f(K-j)$ and $r_{2j} < r_{2j-1}$ (this holds at each stage $2 \leq j \leq K-1$ given the tie-breaking preference of player $2K-1$). The utility then to player $2j-1$ of playing $r_{2j-1} = f(K-j)$ is

$$\begin{aligned} u_{2j-1} &= \big(f(K-j+1) - f(K-j)\big)(K-j) && (8.6)\\ &= \big(f(K-j+1) - f(K-j-1)\big)(K-j-1) && (8.7)\\ &= (K-j)! && (8.8) \end{aligned}$$

where the second equality holds by definition of function f (Eq. (8.4)) and the third equality holds because $(f(K) - f(K-2))(K-2) = (f(K) - f(K-1) + f(K-$

---

[8] On the other hand, on complete d-ary trees, it may be shown that the function $f(k) = \Theta(k) = \Theta(\log_d n)$ for $d \geq 2$ since the number of players, and hence δ_i, grows exponentially with depth K. These growth results on the line graph and the tree seem parallel to the result of Kleinberg and Raghavan [97] (and the elaboration in [25]) which states that the reward required by the root player in order to find an answer to a query with constant probability grows exponentially with the depth of the tree when the branching factor of the tree is $1 < b < 2$, i.e., when each player has an expected number of offsprings $1 < b < 2$, while it grows logarithmically for $b > 2$.

$1) - f(K-2))(K-2) = ((K-2)! + (K-3)!)(K-2) = (K-1)!$. Given the strategies of the rest of the players, player $2j-1$ may not deviate profitably, i.e., $u_{2j-1}(f(K-j), s^*_{-(2j-1)}) \geq u_{2j-1}(r', s^*_{-(2j-1)}), \forall r' \neq f(K-j)$. This is simply because playing an $r' > f(K-j)$ will strictly decrease u_{2j-1} since δ_{2j-1} is already maximized ($\delta_{2j-1} = K-j$ in this case), while playing $r' < f(K-j)$ can at best yield player $2j-1$ the same utility when $r' = f(K-j-1)$ (Eq. (8.7)). The same reasoning holds for player $2j$ who may not deviate profitably by playing $r'' \neq f(K-j-1)$. Specifically, any $r'' < f(K-j-1)$ can at best yield player $2j$ the same utility when $r'' = f(K-j-2)$, and in order to compete with player $2j-1$ (and possibly increase δ_{2j}) player $2j$ must play $r'' > r_{2j-1} = f(K-j)$ which violates the decreasing rewards assumption. Hence neither player at stage j may deviate profitably for all $2 \leq j \leq K-1$. It remains to show that players at stage 1 may not deviate profitably. First, player 1 may not deviate profitably using the same argument we used for player $2j-1$ where $j=1$. The utility to player 1 is $u_1(f(K-1), s^*_{-1}) = (K-1)!$. On the other hand, player 2 gets the same utility as player 1 where $u_2(f(K-2), s^*_{-2}) = (f(K) - f(K-2))(K-2) = (K-1)!$. In the same way, player 2 may not deviate profitably since playing any $r'_2 \neq f(K-2)$ may not increase u_2 given s^*_{-2}. More clearly, in order for player 2 to compete with player 1 and possibly increase δ_2 from $K-2$ to $K-1$, player 2 must play an $r'_2 > f(K-1)$ which in the best case yields a utility $u_2(r'_2, s^*_{-2}) = (f(K) - r'_2)(K-1) < (K-1)!$. Hence, neither player 1 nor player 2 may deviate profitably given the strategies of the other players. Finally, the case for player $2K-1$ is trivial. This concludes the proof. □

This result may be interpreted as follows: if the advertiser were to play strategically assuming she has a marginal utility of at least r^*_d and is aiming for a spanning tree (global discoverability), then $r^*_d = f(K)$ is her Nash strategy in the game induced on the K-stage ring, $\forall K > 2$ (given $\mathbf{s}^*$).

We have shown in Lemma 8.1 that the minimum incentive r^*_d on the line (such that there exists an equilibrium spanning tree for the subgame $G(r^*_d)$) as a function of depth K is $r^*_d = f(K)$. We now extend the result to the ring denoting by $f_r(K)$ the growth function for the ring in order to distinguish it from that of the line, $f(K)$.

Corollary 8.1 *On the ring graph, we have* $f_r(k) = f(k)$ *as given by Lemma* 8.1.

Sketch of Proof We have shown in Theorem 8.3 that $\mathbf{s}^*$ is an equilibrium for the subgame $G(r^*_d)$ for $r^*_d = f(K)$ and that the equilibrium is a spanning tree. What remains to show is that $f(K)$ is the minimum incentive required. This follows by isolating the left branch of the ring, which is a line graph that constitutes of player d and all the players with odd identifiers, and using the same argument of Lemma 8.1 on this branch: an $r_d < f(K)$ allows player 1 to move profitably by playing an $r_1 < f(K-1)$ which violates the spanning tree requirement (by definition of f). □

Next, we present an important result which we utilize to extend the subgame equilibrium of Theorem 8.3 and later as well for the general equilibrium in the repeated version of the game.

8.4.4 Competition Rewards

Competition on the general K-stage ring starts between players 1 and 2 who compete for the last player $2K-1$, knowing that the latter chooses the highest reward route breaking ties by going through the left parent. This section aims at answering the following question: What does it take for player 2 (and the players in the right branch of the ring) to be able to compete with player 1 (and the players in the left branch of the ring) given that the left branch of the ring is preferred to the right branch under the tie-breaking preference of player $2K-1$? Formally, for the players $i, 1 \le i \le n$, let $V_{\text{odd}} = \{i \mid i \text{ is odd, and } i \notin \{2K-1\}\}$ be the subset of players i with odd identifiers (players in the left branch of the ring), while $V_{\text{even}} = \{i \mid i \text{ is even}\}$. In an abuse of notation, denote by $\mathbf{s}_{V_{\text{odd}}}$ the strategy profile of all the players in V_{odd} and similarly for $\mathbf{s}_{V_{\text{even}}}$. We are interested in computing a strategy profile $\hat{\mathbf{s}}_{V_{\text{even}}}$ such that $s_i = \min_{r_i' \in \mathbb{Z}^+}(r_i')$, $\forall s_i \in \hat{\mathbf{s}}_{V_{\text{even}}}$ and such that the players in V_{odd} will have no incentive to compete given $\hat{\mathbf{s}}_{V_{\text{even}}}$, given that player 1 is playing $r_1 = x$ (r_d is assumed to be arbitrarily large here). Specifically, we are interested in the strategy $s_2 \in \hat{\mathbf{s}}_{V_{\text{even}}}$ of player 2 which we refer to as $s_2 = g_K(x)$. In other words, in the subgame $G(h^2)$ for $h^2 = (r_d, r_1 = x, r_2 = g_K(x))$ (for $x \ge f(K-1)$), the outcome tree is guaranteed to have a $\delta_2 = K-1$. For example, in the 3-stage ring of Fig. 8.3(ii), we have $g_3(x) = x+1$ or in other words, player 2 must play at least $r_2 = r_1 + 1$ if she is to win over player 5's business and hence compete with player 1. Finding a closed form for $g_K(x)$ is not necessary for the existence results we seek in this chapter. Our goal here is to show that $g_K(x)$ always exists. We show the existence of $g_k(x)$ in the appendix at the end of this chapter. It may be shown that $g_K(x) - x \le (K-2)!$ peaking at $x = f(K-1) + j(K-2)!, \forall j \ge 0$. Specifically, for $x = f(K-1)$, then $g_K(x) = f(K)$ according to Eq. (8.5).

Having defined $\hat{\mathbf{s}}_{V_{\text{even}}}$ and $g_K(x)$, we now proceed to generalize the result of Theorem 8.3 as follows:

Theorem 8.4 *The profile* $\mathbf{s}^*$ *constructed in Theorem* 8.3 *is a Nash equilibrium for the class of subgames* $G(h^1)$ *for* $f(K) \le h^1 = r_d \le f(K+1)$ *on the* K*-stage ring,* $\forall K > 2$.

Proof We need to show that $\mathbf{s}^*$ is a Nash equilibrium for all the histories $f(K) \le h^1 = r_d \le f(K+1)$. Theorem 8.3 presents the proof for the history $h^1 = r_d = f(K)$. To show that $\mathbf{s}^*$ remains an equilibrium for $f(K) < h^1 \le f(K+1)$, we prove that no player may deviate profitably for all these histories. First, we consider the players in V_{odd} and we note that every player $i \in V_{\text{odd}}$ at stage j may not deviate profitably from playing $f(K-j)$ given s^*_{-i} by definition of f and since δ_i is already maximized ($\delta_i = K-j$). As for the players in V_{even}, we start with player 2. We have shown in Sect. 8.4.4 that in order for player 2 to compete with player 1 (who is playing $r_1 = f(K-1)$) and possibly increase δ_2 (from $K-2$ to $K-1$), player 2 must play $r_2 = g_K(f(K-1)) = f(K)$. Any action $r_2' < r_2$ will not provide enough incentive for the players downstream of 2 to compete and possibly increase δ_2. Hence, the question is to find the value of r_d such that player 2 may not deviate profitably from playing $f(K-2)$ (i.e., not competing) to playing $f(K)$ (i.e., competing). This requirement follows due to the fact that if player 2 has an incentive to compete, then the outcome of the game will oscillate between competing or not (i.e., no equilibrium) as we have demonstrated on the simple 3-stage game in Claim 8.1. The requirement may be stated as

$$\begin{aligned} u_2\big(f(K), s_{-2}\big) &\le u_2\big(f(K-2), s_{-2}\big) \\ \Rightarrow \big(r_d - f(K)\big)(K-1) &\le \big(r_d - f(K-2)\big)(K-2) \\ \Rightarrow r_d &\le (K-1)f(K) - (K-2)f(K-2) \\ &= (K-1)\big(f(K-1) + (K-2)!\big) - (K-2)f(K-2) \\ &= f(K) + (K-1)! \\ &= f(K+1) \end{aligned}$$

where the second inequality follows since as we have already mentioned, player 2 wins over the competition by playing $f(K)$; the fourth and the last inequalities follow from Eq. (8.5). Hence, player 2 may not deviate profitably from playing $f(K-2)$ while $r_d \le f(K+1)$. The same holds for the rest of the players $\in V_{\text{even}}$ since their strategies are contingent on player 2's action r_2. Finally, the case for the last player $2K-1$ is trivial, which concludes the proof. □

We have shown earlier in Claim 8.1 that no SPE exists in the 3-stage version of the game ($K=3$) by showing that the class of subgames for $r_d > 5 = f(3+1)$ do not have a Nash equilibrium. Theorem 8.4 explains the significance of the reward $r_d = 5$ where $5 = f(K+1)$ when $K=3$. Hence, the result for $K=3$ conforms to the general result of Theorem 8.4.

Before concluding this section, we construct the Nash equilibrium for the class of subgames $G(h^1)$ for $h^1 = r_d < f(K)$ on the K-stage ring as follows: Recall first that each player i at stage j observes the history h^j before playing and that $r_{\text{next}(R_i)}$ is a component of the history h_j and particularly of the action profile a^{j-1} (for example for any player i at stage 1, $r_{\text{next}(R_i)} = r_d$). For each player i at stage j, let $s_i^*(1) = 0$, and let $s_i^*(h^j) = f(\kappa^* - 1)$ where $\kappa^* = \max_{\kappa<K} \kappa$ satisfying $f(\kappa) \le r_{\text{next}(R_i)}$. Finally, let $s_{2K-1}^* = 0$.

Theorem 8.5 *The profile* $\mathbf{s}^*$ *thus constructed is a Nash equilibrium for the class of subgames* $G(h^1)$ *for* $h^1 = r_d < f(K)$ *on the* K*-stage ring,* $\forall K > 2$.

Proof For $r_d < f(K)$ there is no competition under s^* and for every player i at stage j, $r_{\text{next}(R_i)} < f(K-j+1)$ which could be shown recursively starting with players at stage 1 and moving downwards. We show that no player can move profitably. By construction of the strategy, each player i will first observe the promised reward $r_{\text{next}(R_i)}$, then compute κ^*, then play $f(\kappa^*-1)$. By definition of κ^* and by definition of f, it must be that for every player i, playing $f(\kappa^* - 1)$ will yield an outcome in which $\delta_i = (\kappa^* - 1)$. Additionally, $(r_{\text{next}(R_i)} - f(\kappa^* - 1))(\kappa^* - 1) \geq (r_{\text{next}(R_i)} - r_i')\delta_i', \forall r_i' \neq f(\kappa^* - 1)$. This is because playing $r_i' > f(\kappa^* - 1)$ will still yield a $\delta_i = (\kappa^* - 1)$ while playing $r_i' < f(\kappa^* - 1)$ will yield a $\delta_i' < \kappa^* - 1$ and a weakly lower utility by construction of f i.e., no player may deviate profitably. The case for the last player $2K - 1$ is trivial which concludes the proof. □

We have characterized the Nash equilibria for all possible histories. With Theorems 8.5 and 8.4, we have constructed the Nash equilibria for the class of subgames $G(h^1)$ for $h^1 \leq f(K+1)$ on the general K-stage ring. On the other hand, we have also shown in Claim 8.1 that the subgame $G(h^1)$ for $h^1 > f(K+1)$ does not have a Nash equilibrium. While the static multi-stage game does not exhibit an equilibrium for $r_d > f(K+1)$, an equilibrium could be constructed in a finitely repeated version of the multi-stage game if we add a *convergence* constraint on utility which essentially dis-incentivizes oscillation as discussed next.

8.4.5 The Repeated Game

First, we revisit the concrete example of the 3-stage game for $r_d = 6$ in Fig. 8.4. An important observation in this simple game is that player 1 may guarantee a payoff of 6 by committing to playing the pure strategy $r_1 = 3$ (assuming player 2 knows that) and there is no equilibrium that yields player 1 a higher payoff. In the general setting when $r_d > 5$, it may be shown that player 1 may threaten player 2 by playing $r_1 = \lceil \frac{r_d}{2} \rceil$ since player 2 will have no incentive to deviate from playing $r_2 = 1$ given r_1. In this case, the payoff to player 1 is r_d when r_d is even and $r_d - 1$ when r_d is odd, while the payoff of player 2 is $r_d - 1$.

While the static game is instructive, it fails to capture the repeated dynamics of BGP and the convergence concept introduced earlier. Many recent efforts have focused on modeling the repeated dynamics inherent to the BGP game [22, 75, 108, 144]. The repeated dynamics are critical to determining the outcome of the game. Afergan [22] shows that BGP is not incentive compatible in the repeated version of the game (at a specific Internet interchange) which violates the incentive-compatibility result of Feigenbaum et al. [58] obtained in the one-shot version of the game. In addition, the game-theoretic BGP model of Levin et al. [108] models the BGP convergence game as an infinitely repeated game and is based on the widely celebrated model of Griffin et al. [75]. We extend the basic game described thus far to model the repeated play in addition to strategic price setting.

In the repeated version of the game, after d advertises itself and declares r_d to its direct neighbors, a finite horizon, repeated, multi-stage game begins. Each round of the game is exactly the multi-stage game described earlier in Sect. 8.4.1 and r_d remains unchanged throughout the game. Recall that the multi-stage model is intended to capture the order of play which is based on proximity to d. The multi-stage game is played a finite number of rounds whereby the rounds are intended to capture repeated strategic price setting among the players over time. Before playing at stage k in round t, player i observes the complete history of all the previous rounds she participated in (rounds $0, 1, \dots, t-1$) as well as the history h^k in round t. We denote this complete history up to stage k in round t by $h^{k,t}$, which is an outcome of the game. Implicitly here, d observes the outcome of every round. For example, player 1 plays at stage 1 in round 0 after observing h^1, then plays again at stage 1 in round 1 after observing $h^{1,1}$ which is the complete history/outcome of stage 0 in addition to h^1 in round 1, and so on.

Let M be the number of rounds played in the finite horizon game where M is sufficiently large and is common knowledge. The pure strategy of a player is now a map $\{s_i^t\}_{t=0}^{M-1}$ where s_i^t is player i's strategy at round t which maps every possible complete history $h^{k,t}$ to a feasible action. The key property that may be defined in the repeated model is *convergence*. More clearly, we have specified earlier that a player receives no utility from an oscillating route. The static game fails to capture this concept of route stability since the game is played only once. The repeated play extension on the other hand allows us to more realistically model route convergence and the respective utility. We have defined convergence earlier to be the convergence of the outcome to a stable tree T_d given some strategy profile $\mathbf{s}$. We have shown that given any profile $\mathbf{s}$ the protocol will converge, and indeed in the multi-stage model, it will do so in a single round, i.e., the outcome of each round t is a tree T_d^t. Convergence in the repeated game is the convergence of player strategies to an equilibrium. Players will only be rewarded for their *stable volume* and this is common knowledge, i.e., a player i may claim rewards from her upstream parent j only when i's subtree is stable. As we shall see, it is this convergence requirement which results in an equilibrium emerging. Intuitively, we are saying that given that best-response dynamics lead to oscillation in the static game, convergence emerges by adding a requirement that players only get paid if they converge. Formally, let T_i^t be i's subtree in the outcome tree T_d^t of round t. In an M round game, T_i is stable only when it remains unchanged from some round t_s onwards. The parameter M, number of rounds, represents the patience of the advertiser d. This is how much d is willing to wait for convergence before distributing any rewards to the players. Suppose that the players discount future payoffs with a common discount factor λ. In order to be able to compare the payoffs of the static game to those of the repeated game, the utility of a player i from repeated play of the multi-stage game, known as the *average discounted payoff* [65], is normalized and is given by

$$\hat{u}_i(\mathbf{s}) = \frac{1-\lambda}{1-\lambda^M} \sum_{t=t_s}^{M-1} \lambda^t u_i\left(s_i^t, s_{-i}^t\right) = \frac{\lambda^{t_s} - \lambda^M}{1-\lambda^M} u_i\left(s_i^t, s_{-i}^t\right) \tag{8.9}$$

where $u_i(s_i^t, s_{-i}^t)$ is the per-round payoff which depends solely on the players' actions and the outcome at round t and is given by Eq. (8.1); and $t_s = \min(t)$ s.t. $T_i^t = T_i^{t'}$, $\forall t < t' \le M$; and $\frac{1-\lambda}{1-\lambda^M}$ is simply a normalization factor which equates to $\frac{1}{\sum_{t=0}^{M-1} \lambda^t}$. First, notice how the utility of a player decreases with t_s according to $\frac{\lambda^{t_s} - \lambda^M}{1-\lambda^M}$ (as t_s varies between 0 and $M-1$). Notice also that when T_i^t is stable, it does not necessarily mean that T_j^t is stable $\forall j \in R_i$. We restrict u_i to consider T_i^t only since this is the subtree that i has control over. The reality is that every T_j^t for $j \in R_i$ must be stable before rewards may flow from d towards i.

Consider the following grim trigger strategy s_1^* for player 1 in the 3-stage game of Fig. 8.4: let $s_1^*(1) = 0$, $s_1^*(2) = 1$ (Theorem 8.5), $s_1^*(3) = s_1^*(4) = 2$ (Theorem 8.4), and $s_1^*(r_d)$ for $r_d > 4$ be to play $r_1 = 2$ in round 0 and continue playing 2 until player 2 plays an $r_2 > 2$ after which switch to playing $r_1 = \lceil \frac{r_d}{2} \rceil$ for the rest of the game.[9] Player 2's strategy s_2^* is to always play $s_2^*(r_d) = 1$ for $r_d \geq 2$ and 0 otherwise. Players 3 and 4 will repeatedly compete in every round playing the strategies of the static game. Finally, $s_5^* = 0$.

Theorem 8.6 *The strategy profile* $\mathbf{s}^*$ *is a Nash equilibrium in the* 3*-stage* ($K = 3$) *game (for every history* h^1*).*

Proof We show that $\mathbf{s}^*$ is a Nash equilibrium for every history h^1. The case for $h^1 = r_d < 5$ follows from Theorems 8.5 and 8.4 since by repeatedly playing the Nash strategies of the static game, then $t_s = 0$ and no player may deviate profitably as can be seen from Eq. (8.9) since $\hat{u}_i = u_i, \forall i$.

For $r_d \geq 5$, player 1 is maximizing her average utility $\hat{u}_1(\mathbf{s}^*)$ by playing $r_1 = f(K-1) = 2$ (by definition of f) given that $r_2 = 1$ in every history and hence player 1 may not gain by any deviation. Notice that this is the only history we consider based on player 2's equilibrium strategy as we are constraining our attention to Nash equilibria. What remains is to show that player 2 may not gain by deviating from playing $r_2 = 1$ (i.e., not competing) in any single round while conforming to $r_2 = 1$ in every other round (this is true since player 2 has one of two options when playing in any round: compete, i.e., $r_2 \geq g_K(r_1)$, or do not compete, i.e,. $r_2 = 1$; a strategy s_2 is a combination of these options across the rounds; by showing that competing in any single round, say t, may not benefit player 2, it follows directly as we shall show that competing in any future round $t' > t$ as well may not benefit player 2 given s_1^*). In order for player 2 to possibly increase her utility (by increasing δ_2), she must deviate by playing an $r_2 \geq g_K(r_1) = r_1 + 1$ as defined in Sect. 8.4.4. If player 2 deviates in round 0, then given the threat strategy of player 1, player 2 will strictly decrease her average payoff since the deviation will cause $t_s > 0$ without any possibility of increasing u_2 for any $t > 0$ when player 1 switches to playing $r_1 = \lceil \frac{r_d}{2} \rceil$. This holds since $u_2(r_2, r_1)$ does not increase, i.e., $u_2(r_2', \lceil \frac{r_d}{2} \rceil) < u_2(1, \lceil \frac{r_d}{2} \rceil), \forall r_2' \geq r_1 = \lceil \frac{r_d}{2} \rceil$ (by construction of $r_1 = \lceil \frac{r_d}{2} \rceil$ as the minimum reward such that player 2 may not benefit by competing given r_d). This argument may be extended to every round in which player 2 participates. Finally, notice that players 3 and 4 may not deviate profitably from repeatedly playing the static strategies (the Bertrand competition) given the strategies of the rest of the players which concludes the proof. □

[9] Here player 1 has an advantage over player 2 and is threatening the latter to force a desirable outcome.

For example, in the $r_d = 6$ subgame of Fig. 8.4, at equilibrium players 1 and 2 are expected to play $(r_1, r_2) = (2, 1)$ in every round yielding $(\hat{u}_1, \hat{u}_2) = (8, 5)$ (again here ignoring the first term of Eq. (8.1)). In the general setting when $r_d \geq 3$, the equilibrium action profile remains $(r_1, r_2) = (2, 1)$ in every round yielding $(\hat{u}_1, \hat{u}_2) = (2r_d - 4, r_d - 1)$. In summary, while no Nash equilibrium exists in the static game, an equilibrium emerges in the repeated model mainly due to the convergence restriction on the players' payoffs which essentially restricts player 2 to concede and avoid oscillation. However, in order for this equilibrium to emerge, player 2 must be aware of the threat strategy of player 1. The result for the 3-stage repeated game on the ring may be extended to the K-stage repeated game.

8.5 Related Work

The Simple Path Vector Protocol (SPVP) formalism [75] develops sufficient conditions for the outcome of a path vector protocol to be stable. The two main components of the formalism are *permitted paths* and local strict *preference* relations over alternate paths to some destination. A respective game-theoretic model was developed by Levin [108] that captures these conditions in addition to incentives in a game theoretic setting. Feigenbaum et al. study incentive issues in BGP by considering least cost path (LCP) policies [58] and more general policies [59]. We have elaborated on the ROUTING-GAME of [59] earlier in Sect. 7.2.1. The model of this chapter is fundamentally different from [58] (and other works based on mechanism design [167]) in that the prices are strategic, the incentive structure is different, and we do not assume the existence of a central "designer" (or bank) that allocates payments to the players but is rather completely distributed as in real markets. The bank assumption is limiting, and an important question posed in [59] is whether the bank can be eliminated and replaced by direct payments by the nodes. A desirable property of our model is that payments are bilateral and may only flow between neighbors where a player i should not be able to send a payment to another player j unless the latter is a direct neighbor. This renders the model more robust to manipulation.

Li et al. [109] study an incentive model for query relaying in peer-to-peer (p2p) networks based on rewards, upon which Kleinberg et al. [97] build to model a more general class of trees. As discussed in the QUERY-GAME earlier in Sect. 7.2.1, Kleinberg and Raghavan [97] allude to a similar version of our distribution game in the context of query incentive networks. They pose the general question of whether an equilibrium exists for general directed acyclic graphs in the query propagation game. Both of these probabilistic models do not account for competition. While we borrow the basic idea, we address a different problem, which is that of route distribution versus information seeking.

Finally, our work relates to price determination in network markets with intermediaries (refer to the work by Blume et al. [36] and the references therein). We have discussed the TRADE-GAME of [36] earlier in Sect. 7.2.1. A main differentiator of this class of work from other work on market pricing is its consideration of intermediaries and the emergence of prices as a result of strategic behavior rather than competitive analysis or truthful mechanisms. Our work specifically involves the cascading of traders on complex network structures.

8.6 Discussion and Conclusion

The Nash equilibria constructed in this chapter are not unique. It is additionally well known that in a multi-stage game setting, the Nash equilibrium notion might not be "credible" as it could present suboptimal responses to histories that would not occur under the equilibrium profile [65], rendering subgame-perfect equilibria more suitable in such circumstances. All the Nash equilibria that we have constructed are credible and are consistent with backward induction for the respective histories of the subgames studied. A distinct aspect of our game is that a player i at stage k may not carry an empty threat to an upstream parent at stage $k-1$, since player i's actions are constrained by the parent's action as dictated by the network structure and the decreasing rewards assumption. In this chapter, we have studied the equilibria existence question only. Other important questions include quantifying how hard is it to find the equilibria, and devising mechanisms to get to them. These questions, in addition to extending the results to general network structures and relaxing the fixed cost assumption, are interesting open problems.

While the distributed incentive model has advantages over centralized mechanisms that rely on a "designer", the model might suffer from exponential growth of rewards which could potentially make it infeasible for sparse and large diameter networks. Quantifying the suitability of this model to general network structures and to the Internet connectivity graph specifically requires further investigation. Interestingly, while it is a complex network, the Internet's connectivity graph is a *small-world* network, i.e., the average distance between any two nodes on the Internet is small [31].

Finally, we have only considered the setting in which d's marginal utility is constant which seems intuitive in a BGP setting where global reachability is the goal, since every node in the DFZ must keep state information about d or else the latter will be unreachable from some parts of the network. Other economic models that assume the network is a market with elastic demand (based on d's utility) and that determine prices based on demand and supply, are interesting to investigate. They may even be more intuitive in settings where it makes sense to advertise (or sell) a piece of information to a local neighborhood.

Appendix: Existence of $g_K(x)$

It is straightforward to show that $g_3(x) = x + 1$ given the Bertrand competition of players 3 and 4 on the 3-stage ring. For $K \geq 4$ and for any $r_1 = x \geq f(K-1)$,[10]

[10]When $x < f(K-1)$, then $g_K(x) = f(K-1)$ by definition of f.

$\hat{\mathbf{s}}_{V_{\text{even}}}$ is part of the solution to the following Integer Linear Program (ILP):[11]

$$\text{minimize} \sum_{\substack{i=2 \\ i \text{ odd}}}^{2K-5} r_i - \beta \sum_{\substack{i=2 \\ i \text{even}}}^{2K-5} r_i$$

s.t.

$$-r_{2K-5} + r_{2K-4} = 1,$$

$$\forall 3 \le i \le 2k-4,$$

$$(r_{i-2} - r_i)(k - l(i)) \ge (r_{i-2} - f(k - l(i) - 1))(k - l(i) - 1),$$

$$\forall 1 \le j \le k-3, \quad r_{2j} \ge r_{2j-1} + 1$$

where β is a sufficiently large constant. The variables in the ILP above signify the actions of the players in the subgame $G(h^2)$ while the constraints guarantee that all players compete while they have an incentive to do so knowing that each player may choose between competing or not. The constraints are constructed based on the definition of $\hat{\mathbf{s}}_{V_{\text{even}}}$ to make sure that players in V_{odd} have no incentive to compete.

[11] Here $g_K(x)$ is the r_2 element in the solution.

References

1. The AKARI architecture design project. http://akari-project.nict.go.jp
2. The alloy analyzer. http://alloy.mit.edu/
3. DARPA active nets. http://www.sds.lcs.mit.edu/darpa-activenet/
4. DARPA network challenge. https://networkchallenge.darpa.mil
5. Delay tolerant networking research group. http://www.dtnrg.org/
6. European future internet portal. http://www.future-internet.eu/
7. GENI: Global environment for network innovations. http://www.geni.org/
8. The handle system. http://www.handle.net
9. NSF future internet architecture project. http://www.nets-fia.net/
10. NSF nets FIND initiative. http://www.nets-find.net/
11. NSF nets FIND projects. http://www.nets-find.net/projects.php
12. 2002 DARPA Active Networks Conference and Exposition (DANCE 2002), 29–31 May 2002, San Francisco, CA, USA. IEEE Computer Society (2002)
13. The internet is broken. MIT Technology Review. http://www.technologyreview.com/article/16356/ (2005)
14. Akamai technologies. http://www.akamai.com (2008)
15. Mobilityfirst. http://mobilityfirst.winlab.rutgers.edu/ (2010)
16. Named Data Networking (NDN) project. http://www.named-data.org (2010)
17. Abbate, J.: Inventing the Internet. MIT Press, Cambridge (2000)
18. Abraham, I., Badola, A., Bickson, D., Malkhi, D., Maloo, S., Ron, S.: Practical locality-awareness for large scale information sharing. In: IPTPS '05: Proceeding of the 4th International Workshop on Peer-to-Peer Systems. Lecture Notes in Computer Science, vol. 3640, pp. 173–181. Springer, Berlin (2005)
19. Abraham, I., Gavoille, C., Malkhi, D., Nisan, N., Thorup, M.: Compact name-independent routing with minimum stretch. In: ACM SPAA '04, pp. 20–24. ACM, New York (2004)
20. Adjie-Winoto, W., Schwartz, E., Balakrishnan, H., Lilley, J.: The design and implementation of an intentional naming system. SIGOPS Oper. Syst. Rev. **34**(2), 22 (2000)
21. Afek, Y., Gafni, E., Ricklin, M.: Upper and lower bounds for routing schemes in dynamic networks. In: 30th Annual Symposium on Foundations of Computer Science, 30 October–1 November 1989, pp. 370–375 (1989)
22. Afergan, M.: Using repeated games to design incentive-based routing systems. In: INFOCOM 2006, pp. 1–13 (2006)
23. Akyildiz, I.F., Akan, O.B., Chen, C., Fang, J., Su, W.: Interplanetary internet: state-of-the-art and research challenges. Comput. Netw. **43**, 75–112 (2003)
24. Andersen, D.G., Balakrishnan, H., Kaashoek, M.F., Morris, R.: Resilient overlay networks. In: Symposium on Operating Systems Principles, pp. 131–145 (2001)
25. Arcaute, E., Kirsch, A., Kumar, R., Liben-Nowell, D., Vassilvitskii, S.: On threshold behav-

J.S. Khoury, C.T. Abdallah, *Internet Naming and Discovery*,
Signals and Communication Technology, DOI 10.1007/978-1-4471-4552-3,

ior in query incentive networks. In: EC '07: Proceedings of the 8th ACM Conference on Electronic Commerce, pp. 66–74. ACM, New York (2007)
26. Arias, M., Cowen, L.J., Laing, K.A., Rajaraman, R., Taka, O.: Compact routing with name independence. In: ACM SPAA '03, pp. 184–192. ACM, New York (2003)
27. Awerbuch, B., Bar-Noy, A., Linial, N., Peleg, D.: Compact distributed data structures for adaptive routing. In: ACM STOC '89, pp. 479–489. ACM, New York (1989)
28. Backstrom, L., Boldi, P., Rosa, M., Ugander, J., Vigna, S.: Four degrees of separation. CoRR abs/1111.4570 (2011)
29. Balakrishnan, H., Lakshminarayanan, K., Ratnasamy, S., Shenker, S., Stoica, I., Walfish, M.: A layered naming architecture for the internet. In: Proceedings of ACM SIGCOMM 2004, pp. 343–352. ACM, Portland (2004)
30. Ballintijn, G., van Steen, M., Tanenbaum, A.S.: Scalable human-friendly resource names. Internet Comput. **5**(5), 20–27 (2001)
31. Barabasi, A.-L.: Linked. Perseus, New York (2002)
32. Bellovin, S.M., Clark, D.D., Perrig, A., Song, D.: A clean-slate design for the next-generation secure internet (2005). This is the report of an NSF workshop held in July, 2005
33. Bent, L., Voelker, G.M.: Whole page performance. In: Workshop on Web Content Caching and Distribution, Boulder CO (2002)
34. Bhattacharjee, B., Calvert, K., Griffioen, J., Spring, N., Sterbenz, J.: Postmodern internetwork architecture. NSF Nets FIND Initiative
35. Blake, S., Black, D., Carlson, M., Davies, E., Wang, Z., Weiss, W.: An architecture for differentiated service. RFC 2475 (1998)
36. Blume, L., Easley, D., Kleinberg, J., Tardos, E.: Trading networks with price-setting agents. In: EC '07, pp. 143–151. ACM, New York (2007)
37. Blumenthal, M.S., Clark, D.D.: Rethinking the design of the internet: the end-to-end arguments vs. the brave new world. In: Communications Policy in Transition: The Internet and Beyond, pp. 91–139 (2001)
38. Boguna, M., Papadopoulos, F., Krioukov, D.: Sustaining the internet with hyperbolic mapping. Nat. Commun. **1**, 62 (2010)
39. Boon Thau Loo Jonathan Smith, Z.I.: Wireless knowledge infrastructure (wiki). NSF Nets FIND Initiative
40. Braden, R., Faber, T., Handley, M.: From protocol stack to protocol heap: role-based architecture. SIGCOMM Comput. Commun. Rev. **33**(1), 17–22 (2003)
41. Brady, A., Cowen, L.: Compact routing on power-law graphs with additive stretch. ALENEX (2006)
42. Caesar, M., Condie, T., Kannan, J., Lakshminarayanan, K., Stoica, I.: Rofl: routing on flat labels. In: Proceedings of SIGCOMM 2006, pp. 363–374. ACM, New York (2006)
43. Caesar, M., Rexford, J.: Bgp routing policies in isp networks. IEEE Netw. **19**(6), 5–11 (2005)
44. Campbell, A.T., Meer, H.G.D., Kounavis, M.E., Miki, K., Vicente, J.B., Villela, D.: A survey of programmable networks. SIGCOMM Comput. Commun. Rev. **29**(2), 7–23 (1999)
45. Castineyra, I., Chiappa, N., Steenstrup, M.: RFC 1992: the nimrod routing architecture (1996)
46. Cerf, V., Kahn, R.: A protocol for packet network interconnection. IEEE Trans. Commun. **22**(5), 637–648 (1974)
47. Chiappa, J.N.: Endpoints and endpoint names: a proposed enhancement to the internet architecture. http://www.chiappa.net/tech/endpoints.txt (1999)
48. Cisco Systems Inc.: Internetworking Technologies Handbook. Cisco, Indianapolis (2004)
49. Clark, D.: The design philosophy of the darpa internet protocols. In: Proceedings of SIGCOMM 1988, pp. 106–114. ACM, New York (1988)
50. Clark, D., Braden, R., Falk, A., Pingali, V.: Fara: reorganizing the addressing architecture. In: FDNA '03: ACM Workshop on Future Directions in Network Architecture, pp. 313–321. ACM, New York (2003)
51. Clark, D.D., Partridge, C., Ramming, J.C., Wroclawski, J.T.: A knowledge plane for the internet. In: SIGCOMM '03: Proceedings of the 2003 Conference on Applications, Tech-

nologies, Architectures, and Protocols for Computer Communications, pp. 3–10. ACM, New York (2003)
52. Coates, M., Hero, A., Nowak, R., Yu, B.: Internet tomography. IEEE Signal Process. Mag. **19**, 47–65 (2002)
53. Coulouris, G.F., Dollimore, J.: Distributed Systems: Concepts and Design. Addison-Wesley Longman, Boston (1988)
54. Crowcroft, J., Hand, S., Mortier, R., Roscoe, T., Warfield, A.: Plutarch: an argument for network pluralism. SIGCOMM Comput. Commun. Rev. **33**(4), 258–266 (2003)
55. Easley, D., Kleinberg, J.: Networks, Crowds, and Markets: Reasoning About a Highly Connected World. Cambridge University Press, Cambridge (2010). ISBN:9780521195331
56. Faloutsos, M., Faloutsos, P., Faloutsos, C.: On power-law relationships of the internet topology. In: SIGCOMM '99: Proceedings of the Conference on Applications, Technologies, Architectures, and Protocols for Computer Communication, pp. 251–262. ACM, New York (1999)
57. Feigenbaum, J., Papadimitriou, C., Sami, R., Shenker, S.: A BGP-based mechanism for lowest-cost routing. Distrib. Comput. **18**(1), 61–72 (2005)
58. Feigenbaum, J., Papadimitriou, C., Sami, R., Shenker, S.: A BGP-based mechanism for lowest-cost routing. Distrib. Comput. **18**(1), 61–72 (2005)
59. Feigenbaum, J., Ramachandran, V., Schapira, M.: Incentive-compatible interdomain routing. In: EC '06, pp. 130–139. ACM, New York (2006)
60. Feigenbaum, J., Shenker, S.: Distributed algorithmic mechanism design: Recent results and future directions. In: In Proceedings of the 6th International Workshop on Discrete Algorithms and Methods for Mobile Computing and Communications, pp. 1–13. ACM, New York (2002)
61. Feldmann, A.: Internet clean-slate design: what and why? SIGCOMM Comput. Commun. Rev. **37**(3), 59–64 (2007)
62. Floyd, S., Paxson, V.: Difficulties in simulating the internet. IEEE/ACM Trans. Netw. **9**(4), 392–403 (2001)
63. Ford, B., Strauss, J., Lesniewski-Laas, C., Rhea, S., Kaashoek, F., Morris, R.: Persistent personal names for globally connected mobile devices. In: Proceedings of the 7th USENIX Symposium on Operating Systems Design and Implementation (OSDI '06), Seattle, WA (2006)
64. Fraigniaud, P., Gavoille, C.: Routing in trees. In: ICALP '01: Proceedings of the 28th International Colloquium on Automata, Languages and Programming, pp. 757–772. Springer, London (2001)
65. Fudenberg, D., Tirole, J.: Game Theory. MIT Press, Cambridge (1991)
66. Fuller, V., Meyer, D., Farinacci, D.: LISP alternative topology (LISP+ALT). http://tools.ietf.org/html/draft-fuller-lisp-alt-03.txt
67. Gao, L.: On inferring autonomous system relationships in the internet. IEEE/ACM Trans. Netw. **9**(6), 733–745 (2001)
68. Garlan, D., Allen, R., Ockerbloom, J.: Architectural mismatch, or, why it's hard to build systems out of existing parts. In: Proceedings of the 17th International Conference on Software Engineering, Seattle, WA, pp. 179–185 (1995)
69. Gavoille, C.: Routing in distributed networks: overview and open problems. SIGACT News **32**(1), 36–52 (2001)
70. Gavoille, C., Gengler, M.: Space-efficiency for routing schemes of stretch factor three. J. Parallel Distrib. Comput. **61**(5), 679–687 (2001)
71. Gavoille, C., Peleg, D.: Compact and localized distributed data structures. Distrib. Comput. **16**(2–3), 111–120 (2003)
72. Gavoille, C., Pérennès, S.: Memory requirement for routing in distributed networks. In: Proceedings of the Fifteenth Annual ACM Symposium on Principles of Distributed Computing. PODC '96, pp. 125–133. ACM, New York (1996)
73. Giloi, W.K.: Towards a taxonomy of computer architecture based on the machine data type view. In: ISCA '83: Proceedings of the 10th Annual International Symposium on Computer

Architecture, pp. 6–15. IEEE Computer Society Press, Los Alamitos (1983)
74. Greenhalgh, A., Huici, F., Hoerdt, M., Papadimitriou, P., Handley, M., Mathy, L.: Flow processing and the rise of commodity network hardware. SIGCOMM Comput. Commun. Rev. **39**(2), 20–26 (2009). doi:10.1145/1517480.1517484
75. Griffin, T.G., Shepherd, F.B., Wilfong, G.: Policy disputes in path-vector protocols. In: ICNP '99, p. 21. IEEE Computer Society, Washington (1999)
76. Griffin, T.G., Shepherd, F.B., Wilfong, G.: The stable paths problem and interdomain routing. IEEE/ACM Trans. Netw. **10**(2), 232–243 (2002)
77. Gritter, M., Cheriton, D.R.: An architecture for content routing support in the internet. In: USITS'01: Proceedings of the 3rd Conference on USENIX Symposium on Internet Technologies and Systems, p. 4. USENIX Association, Berkeley (2001)
78. Gruteser, M.: A geometric stack for location-aware networking. NSF Nets FIND Initiative
79. Gummadi, R., Govindan, R., Kothari, N., Karp, B., Kim, Y.-J., Shenker, S.: Reduced state routing in the internet. In: HotNets 04 (2004)
80. Hafner, K., Lyon, M.: Where Wizards Stay up Late: The Origins of the Internet. Simon and Schuster, New York (1996)
81. Hauzeur, B.M.: A model for naming, addressing and routing. ACM Trans. Inf. Syst. **4**(4), 293–311 (1986)
82. Hennessy, J.L., Patterson, D.A.: Computer Architecture: A Quantitative Approach. Morgan Kaufmann, San Francisco (2003)
83. Herrin, W.: What does a BGP route cost? http://bill.herrin.us/network/bgpcost.html (2008)
84. Hicks, M., Moore, J., Alexander, D., Gunter, C., Nettles, S.: PLANet: an active internetwork (1999)
85. Hicks, M.W., Keromytis, A.D., Smith, J.M.: A secure plan. In: DANCE, pp. 224–237 (2002)
86. Horowitz, D., Kamvar, S.: The anatomy of a large scale social search engine. In: WWW, (2010)
87. Huitema, C., Weerahandi, S.: Internet measurements: the rising tide and the dns snag. In: Proceedings of the 13th ITC Specialist Seminar on IP Traffic Measurement Modeling and Management. IPseminar, ITC, Monterrey, CA, USA (2000)
88. Huston, G.: BGP in 2008. http://www.potaroo.net/ispcol/2009-03/bgp2008.html (2008)
89. Jackson, D.: Software Abstractions: Logic, Language, and Analysis. MIT Press, Cambridge (2006). 0262101149
90. Jacobson, V., Smetters, D.K., Thornton, J.D., Plass, M.F., Briggs, N.H., Braynard, R.L.: Networking named content. In: CoNEXT '09: Proceedings of the 5th International Conference on Emerging Networking Experiments and Technologies, pp. 1–12. ACM, New York (2009)
91. Kahn, R., Abdallah, C., Jerez, H., Heileman, G., Shu, W.: The transient network architecture (TNA). NSF Nets FIND Initiative
92. Karsten, M., Keshav, S., Prasad, S., Beg, M.: An axiomatic basis for communication. In: Proceedings of SIGCOMM 2007, pp. 217–228. ACM, New York (2007)
93. Khoury, J., Abdallah, C.T., Crichigno, J.: Incentivizing cooperation in sensor & control networks. In: IEEE MSC'11, Denver, CO (2011)
94. Kleinberg, J.: Complex networks and decentralized search algorithms. In: Proceedings of the International Congress of Mathematicians (ICM) (2006)
95. Kleinberg, J.: The small-world phenomenon: an algorithm perspective. In: STOC '00: Proceedings of the Thirty-Second Annual ACM Symposium on Theory of Computing, pp. 163–170 (2000)
96. Kleinberg, J.: The convergence of social and technological networks. Commun. ACM **51**(11), 66–72 (2008)
97. Kleinberg, J., Raghavan, P.: Query incentive networks. In: FOCS '05, pp. 132–141. IEEE Computer Society, Washington (2005)
98. Kleinberg, R.: Geographic routing using hyperbolic space. In: INFOCOM 2007: 26th IEEE International Conference on Computer Communications, pp. 1902–1909. IEEE, New York (2007)

99. Koponen, T., Chawla, M., Chun, B.-G., Ermolinskiy, A., Kim, K.H., Shenker, S., Stoica, I.: A data-oriented network architecture. In: Proceedings of SIGCOMM'07. ACM, Kyoto (2007)
100. Korman, A.: Compact routing schemes for dynamic trees in the fixed port model. In: ICDCN '09: Proceedings of the 10th International Conference on Distributed Computing and Networking, pp. 218–229. Springer, Berlin (2009)
101. Korman, A., Peleg, D.: Dynamic routing schemes for general graphs. In: ICALP'06: Proceedings of the 33rd International Conference on Automata, Languages and Programming, Part I, pp. 619–630. Springer, Berlin (2006)
102. Krioukov, D., Claffy, K.C., Fall, K., Brady, A.: On compact routing for the internet. SIGCOMM Comput. Commun. Rev. **37**(3), 41–52 (2007)
103. Krioukov, D., Claffy, K., Fall, K.: Greedy routing on hidden metric spaces as a foundation of scalable routing architectures without topology updates. NSF Nets FIND Initiative
104. Kurose, J.F., Ross, K.W.: Computer Networking: A Top-Down Approach Featuring the Internet. Addison-Wesley Longman, New York (2006)
105. Laing, K.A.: Name-independent compact routing in trees. Technical Report, Tufts University [online]. http://www.cs.tufts.edu/tech_reports/reports/2003-2/report.pdf
106. Laing, K.A.: Name-independent compact routing in trees. Inf. Process. Lett. **103**(2), 57–60 (2007)
107. Lakshminarayanan, K., Adkins, D., Perrig, A., Stoica, I.: Securing user-controlled routing infrastructures. IEEE/ACM Trans. Netw. **16**(3), 549–561 (2008)
108. Levin, H., Schapira, M., Zohar, A.: Interdomain routing and games. In: STOC '08, pp. 57–66. ACM, New York (2008)
109. Li, C., Yu, B., Sycara, K.: An incentive mechanism for message relaying in unstructured peer-to-peer systems. In: AAMAS '07, pp. 1–8. ACM, New York (2007)
110. Mahadevan, P., Krioukov, D.V., Fomenkov, M., Dimitropoulos, X.A., Claffy, K.C., Vahdat, A.: The internet as-level topology: three data sources and one definitive metric. Comput. Commun. Rev. **36**(1), 17–26 (2006)
111. Mas-Colell, A., Whinston, M.D., Green, J.R.: Microeconomic Theory. Oxford University Press, Oxford (1995)
112. Massey, D., Wang, L., Zhang, B., Zhang, L.: A scalable routing system design for future internet. In: ACM SIGCOMM Workshop on IPv6 and the Future of the Internet. ACM, New York (2007)
113. Maymounkov, P., Mazières, D.: Kademlia: A peer-to-peer information system based on the xor metric. In: Revised Papers from the First International Workshop on Peer-to-Peer Systems, IPTPS '01, pp. 53–65. Springer, London (2002)
114. Mazières, D., Kaminsky, M., Kaashoek, M.F., Witchel, E.: Separating key management from file system security. In: SOSP '99: Proceedings of the Seventeenth ACM Symposium on Operating Systems Principles, pp. 124–139. ACM, New York (1999)
115. Meyer, D., Zhang, L., Fall, K.: Report from the IAB workshop on routing and addressing. Internet RFC 4984, September 2007
116. Milgram, S.: The small world problem. Psychol. Today **2**, 60–67 (1967)
117. Mockapetris, P.: RFC 1035: domain names implementation and specification (1987)
118. Moskowitz, R., Nikander, P., Jokela, P.: Host identity protocol architecture. RFC 4423 (2006)
119. Ng, T.S.E., Stoica, I., Zhang, H.: A waypoint service approach to connect heterogeneous internet address spaces. In: Proceedings of the General Track: 2002 USENIX Annual Technical Conference, pp. 319–332. USENIX Association, Berkeley (2001)
120. Nikander, P.: Identifier/locator separation: exploration of the design space (ILSE). IETF Internet Draft draft-nikander-ram-ilse-00 (2007)
121. Nikander, P., Arkko, J., Ohlman, B.: Host identity indirection infrastructure (hi3). In: The Second Swedish National Computer Networking Workshop (2004)
122. Nisan, N., Ronen, A.: Algorithmic mechanism design. In: Proceedings of the 31st ACM Symposium on Theory of Computing, Atlanta, GA (1999)

123. Nisan, N., Roughgarden, T., Tardos, E., Vazirani, V.V.: Algorithmic Game Theory. Cambridge University Press, New York (2007). 0521872820
124. ODell, M.: An alternate addressing architecture for IPv6. IETF Draft. February 1997
125. Papadimitriou, C.: Algorithms, games, and the internet. In: STOC '01: Proceedings of the Thirty-Third Annual ACM Symposium on Theory of Computing, pp. 749–753. ACM, New York (2001)
126. Papadopoulos, F., Krioukov, D., Bogua, M., Vahdat, A.: Greedy forwarding in dynamic scale-free networks embedded in hyperbolic metric spaces. In: INFOCOM, 2010 Proceedings IEEE, pp. 1–9 (2010)
127. Partridge, C.: Forty data communications research questions. SIGCOMM Comput. Commun. Rev. **41**(5), 24–35 (2011)
128. Partridge, C., Mendez, T., Milliken, W.: Host anycasting service. RFC 1546 (1993)
129. Perkins, C.E.: RFC 3220: IP mobility support for IPv4 (2002)
130. Ramakrishna, P.F.: Ipnl: a nat-extended internet architecture. In: Proceedings of SIGCOMM 2001, pp. 69–80. ACM, New York (2001)
131. Ratnasamy, S., Francis, P., Handley, M., Karp, R., Shenker, S.: A scalable content-addressable network. In: SIGCOMM'01: Proceedings ACM Interest Group on Data Communication SIGCOMM 2001, ACM, San Diego (2001)
132. Rekhter, Y., Li, T., Hares, S.: RFC 4271: a border gateway protocol 4 (BGP-4) (2006)
133. Rexford, J., Dovrolis, C.: Future internet architecture: clean-slate versus evolutionary research. Commun. ACM **53**, 36–40 (2010)
134. Rhea, S., Godfrey, B., Karp, B., Kubiatowicz, J., Ratnasamy, S., Shenker, S., Stoica, I., Yu, H.: Opendht: a public dht service and its uses. In: Proceedings of SIGCOMM '05, pp. 73–84. ACM, New York (2005)
135. Rosenberg, J., Schulzrinne, H., et al.: RFC 3261: session initiation protocol (2002)
136. Rowstron, A., Druschel, P.: Pastry: Scalable, distributed object location and routing for large-scale peer-to-peer systems. In: Proceedings of IFIP/ACM Middleware 2001. ACM, Heidelberg (2001)
137. Saltzer, J.: On the naming and binding of network destinations. RFC 1498
138. Saltzer, J.H., Reed, D.P., Clark, D.D.: End-to-end arguments in system design. ACM Trans. Comput. Syst. **2**(4), 277–288 (1984)
139. Santoro, N., Khatib, R.: Labelling and implicit routing in networks. Comput. J. **28**(1), 5–8 (1985). doi:10.1093/comjnl/28.1.5
140. Schmid, S., Eggert, L., Brunner, M., Quittek, J.: TurfNet: an architecture for dynamically composable networks. In: WAC, pp. 94–114 (2004)
141. Segaller, S., Broadcasting, O.P.: Nerds 2.0.1: A Brief History of the Internet. TVBooks (1998)
142. Serrano, M.A., Krioukov, D., Boguna, M.: Self-similarity of complex networks and hidden metric spaces. Phys. Rev. Lett. **100**(7), 078701 (2008)
143. Shakkottai, S., Srikant, R.: Network Optimization and Control. Foundations and Trends in Networking, vol. 2. Now Publishers, Boston (2007)
144. Shakkottai, S., Srikant, R.: Economics of network pricing with multiple isps. IEEE/ACM Trans. Netw. **14**(6), 1233–1245 (2006)
145. Shenker, S.: Fundamental design issues for the future internet. IEEE J. Sel. Areas Commun. **13**(7), 1176–1188 (1995)
146. Shoch, J.: Inter-network naming, addressing, and routing. In: IEEE COMPCON, pp. 72–79. IEEE, New York (1978)
147. Snoeren, A.C., Balakrishnan, H.: An end-to-end approach to host mobility. In: Sixth Annual ACM/IEEE International Conference on Mobile Computing and Networking (2000)
148. Stevens, W.R., Fall, K.: TCP/IP Illustrated: The Protocols V. 1. Addison-Wesley, New York (2009)
149. Stoica, I., Adkins, D., Zhuang, S., Shenker, S., Surana, S.: Internet indirection infrastructure. IEEE/ACM Trans. Netw. **12**(2), 205–218 (2004)

150. Stoica, I., Morris, R., Liben-Nowell, D., Karger, D.R., Kaashoek, M.F., Dabek, F., Balakrishnan, H.: Chord: a scalable peer-to-peer lookup protocol for internet applications. IEEE/ACM Trans. Netw. **11**(1), 17–32 (2003)
151. Su, A.-J., Choffnes, D.R., Kuzmanovic, A., Bustamante, F.E.: Drafting behind akamai (travelocity-based detouring). In: SIGCOMM '06: Proceedings of the 2006 Conference on Applications, Technologies, Architectures, and Protocols for Computer Communications, pp. 435–446. ACM, New York (2006)
152. Subramanian, L., Caesar, M., Ee, C.T., Handley, M., Mao, M., Shenker, S., Stoica, I.: Hlp: a next generation inter-domain routing protocol. In: SIGCOMM '05, pp. 13–24. ACM, New York (2005)
153. Sun, S.: Establishing persistent identity using the handle system. In: Tenth International World Wide Web Conference (2001)
154. Tanenbaum, A.S.: Computer Networks. Prentice Hall, New York (2003). ISBN:9780130661029
155. Tennenhouse, D.L., Smith, J.M., Sincoskie, W.D., Wetherall, D.J., Minden, G.J.: A survey of active network research. IEEE Commun. Mag. **35**(1), 80–86 (1997)
156. Tennenhouse, D.L., Wetherall, D.J.: Towards an active network architecture. SIGCOMM Comput. Commun. Rev. **26**(2), 5–17 (1996)
157. Thorup, M., Zwick, U.: Compact routing schemes. In: ACM SPAA '01, pp. 1–10. ACM, New York (2001)
158. Turányi, Z., Valkó, A., Campbell, A.T.: 4+4: an architecture for evolving the internet address space back toward transparency. SIGCOMM Comput. Commun. Rev. **33**(5), 43–54 (2003)
159. Vahdat, A., Dahlin, M., Anderson, T., Aggarwal, A.: Active names: flexible location and transport of wide-area resources. In: USITS'99: Proceedings of the 2nd Conference on USENIX Symposium on Internet Technologies and Systems, vol. 2, p. 14. USENIX Association, Berkeley (1999)
160. Vakali, A., Pallis, G.: Content delivery networks: status and trends. IEEE Internet Comput. **7**(6), 68–74 (2003)
161. Venkataramani, A., Towsley, D.: A swarming architecture for Internet data transfer. NSF Nets FIND Initiative
162. Watts, D.J., Strogatz, S.H.: Collective dynamics of small-world networks. Nature **393**(6684), 440–442 (1998)
163. Weiss, A.: Net neutrality?: There's nothing neutral about it. Networker **10**(2), 18–25 (2006)
164. Wolf, T.: Service-centric end-to-end abstractions for network architecture. NSF Nets FIND Initiative
165. Yang, X.: Nira: a new internet routing architecture. In: FDNA '03: ACM Workshop on Future Directions in Network Architecture, pp. 301–312. ACM, New York (2003)
166. Yates, R., Raychaudhuri, D., Paul, S., Kurose, J.: Postcards from the edge: a cache-and-forward architecture for the future internet. NSF Nets FIND Initiative
167. Yuen, S., Li, B.: Strategyproof mechanisms towards dynamic topology formation in autonomous networks. Mob. Netw. Appl. **10**(6), 961–970 (2005)
168. Zhao, B.Y., Huang, L., Stribling, J., Rhea, S.C., Joseph, A.D., Kubiatowicz, J.D.: Tapestry: a resilient global-scale overlay for service deployment. IEEE J. Sel. Areas Commun. **22**(1), 41–53 (2004)

Index

J.S. Khoury, C.T. Abdallah, *Internet Naming and Discovery*,
Signals and Communication Technology, DOI 10.1007/978-1-4471-4552-3,

MIX
Papier aus verantwortungsvollen Quellen
Paper from responsible sources
FSC® C105338

If you have any concerns about our products,
you can contact us on
ProductSafety@springernature.com

In case Publisher is established outside the EU,
the EU authorized representative is:
Springer Nature Customer Service Center GmbH
Europaplatz 3, 69115 Heidelberg, Germany

Printed by Libri Plureos GmbH
in Hamburg, Germany